D1614202

Operator Theory: Advances and
Applications
Vol. 167

Wavelets, Multiscale Systems and Hypercomplex Analysis

Daniel Alpay
Editor

Birkhäuser Verlag
Basel · Boston · Berlin

Editor:

Daniel Alpay
Department of Mathematics
Ben-Gurion University of the Negev
P.O. Box 653
Beer Sheva 84105
Israel
e-mail: dany@math.bgu.ac.il

2000 Mathematics Subject Classification 28A80, 30G35, 31C45, 42C40, 44A15, 62M40, 65N30

A CIP catalogue record for this book is available from the
Library of Congress, Washington D.C., USA

Bibliographic information published by Die Deutsche Bibliothek
Die Deutsche Bibliothek lists this publication in the Deutsche Nationalbibliografie; detailed bibliographic data is available in the Internet at <http://dnb.ddb.de>.

ISBN 3-7643-7587-6 Birkhäuser Verlag, Basel – Boston – Berlin

Printed on acid-free paper produced from chlorine-free pulp. TCF ∞
Cover design: Heinz Hiltbrunner, Basel
Printed in Germany
ISBN-10: 3-7643-7587-6 e-ISBN: 3-7643-7588-4
ISBN-13: 978-3-7643-7587-4

9 8 7 6 5 4 3 2 1 www.birkhauser.ch

Contents

Quand sur l'Arbre de la Connaissance
une idée est assez mûre, quelle volupté
de s'y insinuer, d'y agir en larve,
et d'en précipiter la chute!

(Cioran, Syllogismes de l'amertume, [11, p. 145])

Editorial Introduction

Daniel Alpay

This volume contains a selection of papers on the topics of Clifford analysis and wavelets and multiscale analysis, the latter being understood in a very wide sense. That these two topics become more and more related is illustrated for instance by the book of Marius Mitrea [19]. The papers considering connections between Clifford analysis and multiscale analysis constitute more than half of the present volume. This is maybe the specificity of this collection of articles, in comparison with, for instance, the volumes [12], [7], [13] or [18].

The theory of wavelets is mathematically rich and has many practical applications. From a mathematical point of view it is fascinating to realize that most, if not all, of the notions arising from the theory of analytic functions in the open unit disk (in another language, the theory of discrete time systems) have counterparts when one replaces the integers by the nodes of a homogeneous tree. For a review of the mathematics involved we recommand the paper of G. Letac [16]. More recently, and motivated by the works of Basseville, Benveniste, Nikoukhah and Willsky (see [6], [8], [5]) the editor of this volume together with Dan Volok showed that one can replace the complex numbers by a C^*-algebra built from the structure of the tree, and defined point evaluations with values in this C^*-algebra and a corresponding "Hardy space" in which Cauchy's formula holds. The point evaluation could be used to define in this context the counterpart of classical notions such as Blaschke factors. See [3], [2]. Applications include for instance the FBI fingerprint database, as explained in [15] and recalled in the introduction of the paper of Duktay and Jorgensen in the present volume, and the JPEG2000 image compression standard.

It is also fascinating to realize that a whole function theory, different from the classical theory of several complex variables, can be developed when (say, in the quaternionic context) one considers the hypercomplex variables and the Fueter polynomials and the Cauchy–Kovalevskaya product, in place of the classical polynomials in three independent variables; see [10], [14]. Still, a lot of inspiration can be drawn from the classical case, as illustrated in [1].

The volume consists of eight papers, and we now review their contents:

Classical theory: The theory of second order stationary processes indexed by the nodes of a tree involves deep tools from harmonic analysis; see [4], [9]. Some of these aspects are considered in the paper of H. Heyer, **Stationary random fields over graphs and related structures**. The author considers in particular Karhunen–type representations for stationary random fields over quotient spaces of various kinds.

Nonlinear aspects: In the paper **Teodorescu transform decomposition of multivector fields on fractal hypersurfaces** R. Abreu-Blaya, J. Bory-Reyes and T. Moreno-García consider Jordan domains with fractal boundaries. Clifford analysis tools play a central role in the arguments. In **Methods from multiscale theory and wavelets applied to nonlinear dynamics** by D. Dutkay and P. Jorgensen some new applications of multiscale analysis are given to a nonlinear setting.

Numerical computational aspects: In the paper **A Hierarchical semi-separable Moore–Penrose equation solver**, Patrick Dewilde and Shivkumar Chandrasekaran consider operators with hierarchical semi-separable (HSS) structure and consider their Moore–Penrose representation. The HSS forms are close to the theory of systems on trees, but here the multiresolution really represents computation states. In the paper **Matrix representations and numerical computations of wavelet multipliers**, M.W. Wong and Hongmei Zhu use Weyl–Heisenberg frames to obtain matrix representations of wavelet multipliers. Numerical examples are presented.

Connections with Clifford analysis: Such connections are studied in the paper **Metric Dependent Clifford Analysis with Applications to Wavelet Analysis** by F. Brackx, N. De Scheppe and F. Sommen and in the paper **Clifford algebra-valued Admissible Wavelets Associated to more than 2-dimensional Euclidean Group with Dilations** by J. Zhao and L. Peng, the authors study continuous Clifford algebra wavelet transforms, and they extend to this case the classical reproducing kernel property of wavelet transforms; see, e.g., [17, p. 73] for the latter.

Connections with operator theory: G. Gustafson, in **noncommutative trigonometry**, gives an account of noncommutative operator geometry and its applications to the theory of wavelets.

References

[1] D. Alpay, M. Shapiro, and D. Volok. Rational hyperholomorphic functions in R^4. *J. Funct. Anal.*, 221(1):122–149, 2005.

[2] D. Alpay and D. Volok. Interpolation et espace de Hardy sur l'arbre dyadique: le cas stationnaire. *C.R. Math. Acad. Sci. Paris*, 336:293–298, 2003.

[3] D. Alpay and D. Volok. Point evaluation and Hardy space on a homogeneous tree. *Integral Equations Operator Theory*, 53:1–22, 2005.

[4] J.P. Arnaud. Stationary processes indexed by a homogeneous tree. *Ann. Probab.*, 22(1):195–218, 1994.

[5] M. Basseville, A. Benveniste, and A. Willsky. Multiscale autoregressive processes. Rapport de Recherche 1206, INRIA, Avril 1990.

[6] M. Basseville, A. Benveniste, and A. Willsky. Multiscale statistical signal processing. In *Wavelets and applications (Marseille, 1989)*, volume 20 of *RMA Res. Notes Appl. Math.*, pages 354–367. Masson, Paris, 1992.

[7] J. Benedetto and A. Zayed, editors. *Sampling, wavelets, and tomography.* Applied and Numerical Harmonic Analysis. Birkhäuser Boston Inc., Boston, MA, 2004.

[8] A. Benveniste, R. Nikoukhah, and A. Willsky. Multiscale system theory. *IEEE Trans. Circuits Systems I Fund. Theory Appl.*, 41(1):2–15, 1994.

[9] W. Bloom and H. Heyer. *Harmonic analysis of probability measures on hypergroups*, volume 20 of *de Gruyter Studies in Mathematics.* Walter de Gruyter & Co., Berlin, 1995.

[10] F. Brackx, R. Delanghe, and F. Sommen. *Clifford analysis*, volume 76. Pitman research notes, 1982.

[11] E.M. Cioran. *Syllogismes de l'amertume.* Collection idées. Gallimard, 1976. First published in 1952.

[12] C.E. D'Attellis and E.M. Fernández-Berdaguer, editors. *Wavelet theory and harmonic analysis in applied sciences.* Applied and Numerical Harmonic Analysis. Birkhäuser Boston Inc., Boston, MA, 1997. Papers from the 1st Latinamerican Conference on Mathematics in Industry and Medicine held in Buenos Aires, November 27–December 1, 1995.

[13] L. Debnath. *Wavelet transforms and their applications.* Birkhäuser Boston Inc., Boston, MA, 2002.

[14] R. Delanghe, F. Sommen, and V. Souček. *Clifford algebra and spinor valued functions*, volume 53 of *Mathematics and its applications.* Kluwer Academic Publishers, 1992.

[15] M.W. Frazier. *An introduction to wavelets through linear algebra.* Undergraduate Texts in Mathematics. Springer-Verlag, New York, 1999.

[16] G. Letac. Problèmes classiques de probabilité sur un couple de Gel′fand. In *Analytical methods in probability theory (Oberwolfach, 1980)*, volume 861 of *Lecture Notes in Math.*, pages 93–120. Springer, Berlin, 1981.

[17] S. Mallat. *Une exploration des signaux en ondelettes.* Les éditions de l'École Polytechnique, 2000.

[18] Y. Meyer, editor. *Wavelets and applications*, volume 20 of *RMA: Research Notes in Applied Mathematics*, Paris, 1992. Masson.

[19] M. Mitrea. *Clifford wavelets, singular integrals, and Hardy spaces*, volume 1575 of *Lecture Notes in Mathematics.* Springer-Verlag, Berlin, 1994.

Daniel Alpay
Department of Mathematics
Ben–Gurion University of the Negev
POB 653
Beer-Sheva 84105, Israel
e-mail: dany@math.bgu.ac.il

Operator Theory:
Advances and Applications, Vol. 167, 1–16

Teodorescu Transform Decomposition of Multivector Fields on Fractal Hypersurfaces

Ricardo Abreu-Blaya, Juan Bory-Reyes and Tania Moreno-García

Abstract. In this paper we consider Jordan domains in real Euclidean spaces of higher dimension which have fractal boundaries. The case of decomposing a Hölder continuous multivector field on the boundary of such domains is obtained in closed form as sum of two Hölder continuous multivector fields harmonically extendable to the domain and to the complement of its closure respectively. The problem is studied making use of the Teodorescu transform and suitable extension of the multivector fields. Finally we establish equivalent condition on a Hölder continuous multivector field on the boundary to be the trace of a harmonic Hölder continuous multivector field on the domain.

Mathematics Subject Classification (2000). Primary: 30G35.

Keywords. Clifford analysis, Fractals, Teodorescu-transform.

1. Introduction

Clifford analysis, a function theory for Clifford algebra-valued functions in $\mathbb{R}^n$ satisfying $\mathcal{D}u = 0$, with $\mathcal{D}$ denoting the Dirac operator, was built in [10] and at the present time becomes an independent mathematical discipline with its own goals and tools. See [12, 18, 19] for more information and further references.

In [2, 3, 17], see also [4, 8] it is shown how Clifford analysis tools offer a lot of new brightness in studying the boundary values of harmonic differential forms in the Hodge-de Rham sense that are in one to one correspondence with the so-called harmonic multivector fields in $\mathbb{R}^n$. In particular, the case was studied of decomposing a Hölder continuous multivector field F_k on a closed (Ahlfors-David) regular hypersurface Γ, which bounds a domain $\Omega \subset \mathbb{R}^n$, as a sum of two Hölder continuous multivector fields $F_k^{\pm}$ harmonically extendable to Ω and $\mathbb{R}^n \setminus (\Omega \cup \Gamma)$ respectively. A set of equivalent assertions was established (see [2], Theorem 4.1) which are of a pure function theoretic nature and which, in some sense, are replacing Condition (A) obtained by Dynkin [13] in studying the analogous problem for harmonic k-forms. For the case of C^∞-hypersurfaces Σ and C^∞-multivector fields

F_k, it is revealed in [3] the equivalence between Dynkin's condition (A) and the so-called conservation law ((CL)-condition)

$$\upsilon\, \partial_{\underline{\omega}} F_k = F_k \partial_{\underline{\omega}}\, \upsilon, \tag{1}$$

where υ denotes the outer unit normal vector on Σ and $\partial_{\underline{\omega}}$ stands for the tangential Dirac operator restrictive to Σ.

In general, a Hölder continuous multivector field F_k admits on Γ the previously mentioned harmonic decomposition if and only if

$$H_\Gamma F_k = F_k H_\Gamma, \tag{2}$$

where H_Γ denotes the Hilbert transform on Γ

$$H_\Gamma u(x) := \frac{2}{A_n} \int_\Sigma \frac{x-y}{|y-x|^n} \upsilon(y)(u(y) - u(x)) d\mathcal{H}^n(y) + u(x)$$

The commutation (2) can be viewed as an integral form of the (CL)-condition. Moreover, a Hölder continuous multivector field F_k is the trace on Γ of a harmonic multivector field in Ω if and only if $H_\Gamma F_k = F_k$. An essential role in proving the above results is played by the Cliffordian Cauchy transform

$$\mathcal{C}_\Sigma u(x) := \frac{1}{A_n} \int_\Sigma \frac{x-y}{|y-x|^n} \upsilon(y) u(y) d\mathcal{H}^n(y)$$

and, in particular, by the Plemelj-Sokhotski formulas, which are valid for Hölder continuous functions on regular hypersurfaces in $\mathbb{R}^n$, see [1, 6].

The question of the existence of the continuous extension of the Clifford–Cauchy transform on a rectifiable hypersurface in $\mathbb{R}^n$ which at the same time satisfies the so-called Ahlfors–David regularity condition is optimally answered in [9].

What would happen with the above-mentioned decomposition for a continuous multivector field if we replace the considered reasonably nice domain with one which has a fractal boundary?

The purpose of this paper is to study the boundary values of harmonic multivector fields and monogenic functions when the hypersurface Γ is a fractal. Fractals appear in many mathematical fields: geometric measure theory, dynamical system, partial differential equations, etc. For example, in [25] some unexpected and intriguing connections has been established between the inverse spectral problem for vibrating fractal strings in $\mathbb{R}^n$ and the famous Riemann hypothesis.

In this paper we treat mainly four kinds of problems: Let Ω be an open bounded domain of $\mathbb{R}^n$ having as a boundary a fractal hypersurface Γ.

I) (Jump Problem) Let u be a Hölder continuous Clifford algebra-valued function on Γ. Under which conditions can one represent u as a sum

$$u = U^+ + U^-, \tag{3}$$

where the functions $U^\pm$ are monogenic in $\Omega_+ := \Omega$ and $\Omega_- := \mathbb{R}^n \setminus \overline{\Omega}$ respectively?

II) Let u be a Hölder continuous Clifford algebra-valued function on Γ. Under which conditions is u the trace on Γ of a monogenic function in $\Omega_\pm$?

III) (Dynkin's type Problem) Let F_k be a Hölder continuous multivector field on Γ. Under which conditions F_k can be decompose as a sum

$$F_k = F_k^+ + F_k^-, \tag{4}$$

where $F_k^\pm$ are harmonic multivector fields in $\Omega_\pm$, respectively?

IV) Let F_k be a Hölder continuous multivector field on Γ. Under which conditions is F_k the trace on Γ of a harmonic multivector field in $\Omega_\pm$?

2. Preliminaries

We thought it to be helpful to recall some well-known, though not necessarily familiar basic properties in Clifford algebras and Clifford analysis and some basic notions about fractal dimensions and Whitney extension theorem as well.

2.1. Clifford algebras and multivectors

Let $\mathbb{R}^{0,n}$ ($n \in \mathbb{N}$) be the real vector space $\mathbb{R}^n$ endowed with a non-degenerate quadratic form of signature $(0, n)$ and let $(e_j)_{j=1}^n$ be a corresponding orthogonal basis for $\mathbb{R}^{0,n}$. Then $\mathbb{R}_{0,n}$, the universal Clifford algebra over $\mathbb{R}^{0,n}$, is a real linear associative algebra with identity such that the elements e_j, $j = 1, \dots, n$, satisfy the basic multiplication rules

$$e_j^2 = -1, \; j = 1, \dots, n;$$

$$e_i e_j + e_j e_i = 0, \; i \neq j.$$

For $A = \{i_1, \dots, i_k\} \subset \{1, \dots, n\}$ with $1 \leq i_1 < i_2 < \cdots < i_k \leq n$, put $e_A = e_{i_1} e_{i_2} \cdots e_{i_k}$, while for $A = \emptyset$, $e_\emptyset = 1$ (the identity element in $\mathbb{R}_{0,n}$). Then $(e_A : A \subset \{1, \dots, n\})$ is a basis for $\mathbb{R}_{0,n}$. For $1 \leq k \leq n$ fixed, the space $\mathbb{R}_{0,n}^{(k)}$ of k vectors or k-grade multivectors in $\mathbb{R}_{0,n}$, is defined by

$$\mathbb{R}_{0,n}^{(k)} = \mathrm{span}_\mathbb{R}(e_A : \; |A| = k).$$

Clearly

$$\mathbb{R}_{0,n} = \sum_{k=0}^{n} \oplus \mathbb{R}_{0,n}^{(k)}.$$

Any element $a \in \mathbb{R}_{0,n}$ may thus be written in a unique way as

$$a = [a]_0 + [a]_1 + \cdots + [a]_n$$

where $[\,]_k : \mathbb{R}_{0,n} \longrightarrow \mathbb{R}_{0,n}^{(k)}$ denotes the projection of $\mathbb{R}_{0,n}$ onto $\mathbb{R}_{0,n}^k$. Notice that for any two vectors x and y, their product is given by

$$xy = x \bullet y + x \wedge y$$

where

$$x \bullet y = \frac{1}{2}(xy + yx) = -\sum_{j=1}^{n} x_j y_j$$

is – up to a minus sign – the standard inner product between x and y, while

$$x \wedge y = \frac{1}{2}(xy - yx) = \sum_{i<j} e_i e_j (x_i y_j - x_j y_i)$$

represents the standard outer product between them.

More generally, for a 1-vector x and a k-vector Y_k, their product xY_k splits into a $(k-1)$-vector and a $(k+1)$-vector, namely:

$$xY_k = [xY_k]_{k-1} + [xY_k]_{k+1}.$$

The inner and outer products between x and Y_k are then defined by

$$x \bullet Y_k = [xY_k]_{k-1} \text{ and } x \wedge Y_k = [xY_k]_{k+1}. \tag{5}$$

For further properties concerning inner and outer products between multivectors, we refer to [12].

2.2. Clifford analysis and harmonic multivector fields

Clifford analysis offers a function theory which is a higher dimensional analogue of the function theory of holomorphic functions of one complex variable.

Consider functions defined in $\mathbb{R}^n, (n > 2)$ and taking values in the Clifford algebra $\mathbb{R}_{0,n}$. Such a function is said to belong to some classical class of functions if each of its components belongs to that class.

In $\mathbb{R}^n$ we consider the Dirac operator:

$$\mathcal{D} := \sum_{j=1}^{n} e_j \partial_{x_j},$$

which plays the role of the Cauchy Riemann operator in complex analysis.

Due to $\mathcal{D}^2 = -\Delta$, where Δ being the Laplacian in $\mathbb{R}^n$ the monogenic functions are harmonic.

Suppose $\Omega \subset \mathbb{R}^n$ is open, then a real differentiable $\mathbb{R}_{0,n}$-valued function f in Ω is called left (right) monogenic in Ω if $\mathcal{D}f = 0$ (resp. $f\mathcal{D} = 0$) in Ω. The notion of left (right) monogenicity in $\mathbb{R}^n$ provides a generalization of the concept of complex analyticity to Clifford analysis.

Many classical theorems from complex analysis could be generalized to higher dimensions by this approach. Good references are [10, 12]. The space of left monogenic functions in Ω is denoted by $\mathcal{M}(\Omega)$. An important example of a two-sided monogenic function is the fundamental solution of the Dirac operator, given by

$$e(x) = \frac{1}{A_n} \frac{\overline{x}}{|x|^n}, \quad x \in \mathbb{R}^n \setminus \{0\}.$$

Hereby A_n stands for the surface area of the unit sphere in $\mathbb{R}^n$. The function $e(x)$ plays the same role in Clifford analysis as the Cauchy kernel does in complex analysis.

Notice that if F_k is a k-vector-valued function, i.e.,

$$F_k = \sum_{|A|=k} e_A F_{k,A}\,,$$

then, by using the inner and outer products (5), the action of $\mathcal{D}$ on F_k is given by

$$\mathcal{D}F_k = \mathcal{D} \bullet F_k + \mathcal{D} \wedge F_k \tag{6}$$

As $\overline{\mathcal{D}F_k} = \overline{F_k}\,\overline{\mathcal{D}}$ with $\overline{\mathcal{D}} = -\mathcal{D}$ and $\overline{F_k} = (-1)^{\frac{k(k+1)}{2}} F_k$, it follows that if F_k is left monogenic, then it is right monogenic as well. Furthermore, in view of (6), F_k is left monogenic in Ω if and only if F_k satisfies in Ω the system of equations

$$\begin{cases} \mathcal{D} \bullet F_k = 0 \\ \mathcal{D} \wedge F_k = 0. \end{cases} \tag{7}$$

A k-vector-valued function F_k satisfying (7) in Ω is called a harmonic (k-grade) multivector field in Ω.

The following lemma will be much useful in what follows.

Lemma 2.1. *Let u be a real differentiable $\mathbb{R}_{0,n}$-valued function in Ω admitting the decomposition*

$$u = \sum_{k=0}^{n} [u]_k.$$

Then u is two-sided monogenic in Ω if and only if $[u]_k$ is a harmonic multivector field in Ω, for each $k = 0, 1, \ldots, n$.

2.3. Fractal dimensions and Whitney extension theorem

Before stating the main result of this subsection we must define the s-dimensional Hausdorff measure. Let $\mathbf{E} \subset \mathbb{R}^n$ then the Hausdorff measure $\mathcal{H}^s(\mathbf{E})$ is defined by

$$\mathcal{H}^s(\mathbf{E}) := \lim_{\delta \to 0} \inf\Big\{\sum_{k=1}^{\infty} (\operatorname{diam} B_k)^s \,:\, \mathbf{E} \subset \cup_k B_k,\ \operatorname{diam} B_k < \delta\Big\},$$

where the infimum is taken over all countable δ-coverings $\{B_k\}$ of $\mathbf{E}$ with open or closed balls. Note that $\mathcal{H}^n$ coincides with the Lebesgue measure $\mathcal{L}^n$ in $\mathbb{R}^n$ up to a positive multiplicative constant.

Let $\mathbf{E}$ be a bounded set in $\mathbb{R}^n$. The Hausdorff dimension of $\mathbf{E}$, denoted by $\mathrm{H}(\mathbf{E})$, is the infimum of the numbers $s \geq 0$ such that $\mathcal{H}^s(\mathbf{E}) < \infty$. For more details concerning the Hausdorff measure and dimension we refer the reader to [14, 15]. Frequently, see [25], the box dimension is more appropriated dimension than the Hausdorff dimension to measure the roughness of $\mathbf{E}$.

Suppose $\mathcal{M}_0$ denotes a grid covering $\mathbb{R}^n$ and consisting of n-dimensional cubes with sides of length 1 and vertices with integer coordinates. The grid $\mathcal{M}_k$ is obtained from $\mathcal{M}_0$ by division of each of the cubes in $\mathcal{M}_0$ into 2^{nk} different cubes with side length 2^{-k}. Denote by $N_k(\mathbf{E})$ the minimum number of cubes of the grid $\mathcal{M}_k$ which have common points with $\mathbf{E}$.

The box dimension (also called upper Minkowski dimension, see [20, 21, 22, 23]) of $\mathbf{E} \subset \mathbb{R}^n$, denoted by $\mathrm{M}(\mathbf{E})$, is defined by

$$\mathrm{M}(\mathbf{E}) = \limsup_{k\to\infty} \frac{\log N_k(\mathbf{E})}{k\log(2)},$$

if the limit exists.

The box dimension and Hausdorff dimension can be equal (e.g., for the so-called $(n-1)$-rectifiable sets, see [16]) although this is not always valid. For sets with topological dimension $n-1$ we have $n-1 \leq \mathrm{H}(\mathbf{E}) \leq \mathrm{M}(\mathbf{E}) \leq n$. If the set $\mathbf{E}$ has $\mathrm{H}(\mathbf{E}) > n-1$, then it is called a fractal set in the sense of Mandelbrot.

Let $\mathbf{E} \subset \mathbb{R}^n$ be closed. Call $\mathcal{C}^{0,\nu}(\mathbf{E}), 0 < \nu < 1$, the class of $\mathbb{R}_{0,n}$-valued functions satisfying on $\mathbf{E}$ the Hölder condition with exponent ν.

Using the properties of the Whitney decomposition of $\mathbb{R}^n \setminus \mathbf{E}$ (see [26], p. 174) the following Whitney extension theorem is obtained.

Theorem 2.1 (Whitney Extension Theorem). *Let $u \in \mathcal{C}^{0,\nu}(\mathbf{E}), 0 < \nu < 1$. Then there exists a compactly supported function $\tilde{u} \in \mathcal{C}^{0,\nu}(\mathbb{R}^n)$ satisfying*

(i) $\tilde{u}|_{\mathbf{E}} = u|_{\mathbf{E}}$,
(ii) $\tilde{u} \in C_\infty(\mathbb{R}^n \setminus \mathbf{E})$,
(iii) $|\mathcal{D}\tilde{u}(x)| \leq C\, dist(x, \mathbf{E})^{\nu-1}$ *for* $x \in \mathbb{R}^n \setminus \mathbf{E}$

3. Jump problem and monogenic extensions

In this section, we derive the solution of the jump problem as well as the problem of monogenically extension of a Hölder continuous functions in Jordan domains. We shall work with fractal boundary data. Recall that $\Omega \subset \mathbb{R}^n$ is called a Jordan domain (see [21]) if it is a bounded oriented connected open set whose boundary is a compact topological hypersurface.

In [22, 23] Kats presented a new method for solving the jump problem, which does not use contour integration and can thus be used on nonrectifiable and fractal curves. A natural multidimensional analogue of such method was adapted immediately within Quaternionic Analysis in [1]. Continuing along the same lines a possible generalization for $n > 2$ could also be envisaged. The jump problem (3) in this situation was considered by the authors in [6, 7]. They were able to show that for $u \in C^{0,\nu}(\Gamma)$, when the Hölder exponent ν and the box dimension $\mathbf{m} := \mathrm{M}(\Gamma)$ of the surface Γ satisfy the relation

$$\nu > \frac{\mathbf{m}}{n}, \tag{8}$$

then a solution of (3) can be given by the formulas

$$U^+(x) = \tilde{u}(x) + T_\Omega \mathcal{D}\tilde{u}(x), \qquad U^-(x) = T_\Omega \mathcal{D}\tilde{u}(x)$$

where $\mathcal{T}_\Omega$ is the Teodorescu transform defined by

$$\mathcal{T}_\Omega v(x) = -\int_\Omega e(y-x)v(y)d\mathcal{L}^n(y).$$

For details regarding the basic properties of Teoderescu transform, we refer the reader to [19].

It is essential here that under condition (8) $\mathcal{D}(\tilde{u}|_\Omega)$ is integrable in Ω with any degree not exceeding $\frac{n-m}{1-\nu}$ (see [7]), i.e., under condition (8) it is integrable with certain exponent exceeding n. At the same time, $\mathcal{T}_\Omega \mathcal{D}\tilde{u}$ satisfies the Hölder condition with exponent $1-\frac{n}{p}$. Therefore we have $\mathcal{T}_\Omega \mathcal{D}\tilde{u} \in C^{0,\mu}(\mathbb{R}^n)$ with

$$\mu < \frac{n\nu - \mathbf{m}}{n - \mathbf{m}}.$$

When condition (8) is violated then some obstructions can be constructed as we shall see through an example in Section 5.

In the sequel we assume ν and $\mathbf{m}$ to be connected by (8). We now return to the problem (II) of extending monogenically a Hölder continuous $\mathbb{R}_{0,n}$-valued function.

For the remainder of this paper let $\Omega := \Omega_+$ be a Jordan domain in $\mathbb{R}^n$ with boundary $\Gamma := \partial\Omega_+$, by Ω_- we denote the complement domain of $\Omega_+ \cup \Gamma$.

Let $u \in C^{0,\nu}(\Gamma)$. If Γ is some kind of reasonably nice boundary, then the classical condition $H_\Gamma u = u$ ($H_\Gamma u = -u$) is sufficient (and in this case also necessary!) for u to be the trace of a left monogenic function in Ω_+ (Ω_-). Of course, if Γ is fractal, then the operator H_Γ loses its usual meaning but the problem (II) remains still meaningful. The key idea here is to indicate how the Hilbert transform could be replaced by the restriction to Γ of the Teodorescu transform so that the letter plays similar role of the former $\mathcal{T}_{\Omega_+}$, but the restriction to Γ of the last one is needed. Obviously our situation is much more general and seems to have been overlooked by many authors.

Theorem 3.1. *Let $u \in C^{0,\nu}(\Gamma)$. If u is the trace of a function $U \in C^{0,\nu}(\Omega_+ \cup \Gamma) \cap \mathcal{M}(\Omega_+)$, then*

$$\mathcal{T}_{\Omega_+} \mathcal{D}\tilde{u}|_\Gamma = 0 \tag{9}$$

Conversely, if (9) is satisfied, then u is the trace of a function $U \in C^{0,\mu}(\Omega_+ \cup \Gamma) \cap \mathcal{M}(\Omega_+)$, for some $\mu < \nu$.

Proof. The proof of the first part involves a mild modification of the proof of lemma 1 in [24]. Let $U^* = \tilde{u} - U$. Let Q_k be the union of cubes of the net $\mathcal{M}_k$ intersecting Γ. Denote $\Omega_k = \Omega_+ \setminus Q_k$, $\Delta_k = \Omega_+ \setminus \Omega_k$ and denote by Γ_k the boundary of Ω_k.

By the definition of $\mathbf{m}$, for any $\epsilon > 0$ there is a constant $C(\epsilon)$ such that $N_k(\Gamma) \le C(\epsilon)2^{k(\mathbf{m}+\epsilon)}$. Then,

$$\mathcal{H}^{n-1}(\Gamma_k) \le 2nC(\epsilon)2^{k(\mathbf{m}-n+1+\epsilon)}$$

As usual, the letters C, C_1, C_2 stand for absolute constants. Since $U^* \in C^{0,\nu}(\Gamma)$, $U^*|_\Gamma = 0$ and any point of Γ_k is distant from Γ by no more than $C_1 \cdot 2^{-k}$, then we have

$$\max_{y\in\Gamma_k} |U^*(y)| \leq C_2\, 2^{-\nu k}.$$

Consequently, for $x \in \Omega_-$, $s = dist(x, \Gamma)$

$$|\int_{\Gamma_k} e(y-x)\upsilon(y)U^*(y)d\mathcal{H}^{n-1}(y)| \leq \frac{2n}{s^{n-1}} C_2 2^{-\nu k} C(\epsilon) 2^{k(\mathbf{m}-n+1+\epsilon)}$$

$$- C_2 C(\epsilon)\frac{2n}{s^{n-1}} 2^{k(\mathbf{m}-n+1-\nu+\epsilon)}.$$

Under condition (8) the right-hand side of the above inequality tends to zero as $k \to \infty$. By Stokes formula we have

$$\int_\Omega e(y-x)\mathcal{D}U^*(y)d\mathcal{L}^n(y) = \lim_{k\to\infty} (\int_{\Delta_k} + \int_{\Omega_k}) e(y-x)\mathcal{D}U^*(y)d\mathcal{L}^n(y)$$

$$= \lim_{k\to\infty} (\int_{\Delta_k} e(y-x)\mathcal{D}U^*(y)d\mathcal{L}^n(y) - \int_{\Gamma_k} e(y-x)\upsilon(y)U^*(y)d\mathcal{H}^{n-1}(y)) = 0.$$

Therefore

$$\mathcal{T}_{\Omega_+}\mathcal{D}\tilde{u}|_\Gamma = \mathcal{T}_{\Omega_+}\mathcal{D}U|_\Gamma = 0.$$

The second assertion follows directly by taking $U = \tilde{u} + \mathcal{T}_{\Omega_+}\mathcal{D}\tilde{u}$. □

For Ω_- the following analogous result can be obtained.

Theorem 3.2. *Let $u \in C^{0,\nu}(\Gamma)$. If u is the trace of a function $U \in C^{0,\nu}(\Omega_- \cup \Gamma) \cap \mathcal{M}(\Omega_-)$, and $U(\infty) = 0$, then*

$$\mathcal{T}_{\Omega_+}\mathcal{D}\tilde{u}|_\Gamma = -u(x) \tag{10}$$

Conversely, if (9) is satisfied, then u is the trace of a function $U \in C^{0,\mu}(\Omega_- \cup \Gamma) \cap \mathcal{M}(\Omega_-)$ and $U(\infty) = 0$, for some $\mu < \nu$.

Remark 3.1. *To be more precise, under condition (9) (resp. (10)) the function $\tilde{u} + \mathcal{T}_{\Omega_+}\mathcal{D}\tilde{u}$ ($\mathcal{T}_{\Omega_+}\mathcal{D}\tilde{u}$) is a monogenic extension of u to Ω_+ (Ω_-) which belongs to $C^{0,\mu}(\Omega_+ \cup \Gamma)$ (resp. $C^{0,\mu}(\Omega_- \cup \Gamma)$) for any*

$$\mu < \frac{n\nu - \mathbf{m}}{n - \mathbf{m}}.$$

On the other hand, one can state quite analogous results for the case of considering (right) monogenic extensions. For that it is only necessary to replace in both conditions (9) and (10) $\mathcal{T}_{\Omega_+}\mathcal{D}\tilde{u}|_\Gamma$ by $\tilde{u}\mathcal{D}\mathcal{T}_{\Omega_+}|_\Gamma$, where for a $\mathbb{R}_{0,n}$-valued function f

$$f\mathcal{T}_\Omega(x) := -\int_\Omega f(y)e(y-x)d\mathcal{L}^n(y).$$

Theorem 3.3. *If $U \in C^{0,\nu}(\Omega_+ \cup \Gamma) \cap \mathcal{M}(\Omega_+)$ has trace $u := U|_\Gamma$. Then the following assertions are equivalent:*

(i) *U is two-sided monogenic in Ω_+*
(ii) $\mathcal{T}_{\Omega_+}\mathcal{D}\tilde{u}|_\Gamma = \tilde{u}\mathcal{D}\mathcal{T}_{\Omega_+}|_\Gamma$

Proof. That (i) implies (ii) follows directly from Theorem 3.1 and its right-hand side version. Assume (ii) is satisfied. Since U is (left) monogenic in Ω_+, then by Theorem 3.1 we have

$$\mathcal{T}_{\Omega_+}\mathcal{D}\tilde{u}|_\Gamma = \tilde{u}\mathcal{D}\mathcal{T}_{\Omega_+}|_\Gamma = 0.$$

The second equality implies the existence of a function U^* which is (right) monogenic in Ω_+ and such that $U^*|_\Gamma = u$. Put $w = U - U^*$, then $\Delta w = 0$ in Ω_+ and $w|_\Gamma = 0$. By the classical Dirichlet problem we conclude that $U = U^*$ in Ω_+, which means that U is two-sided monogenic in Ω_+. □

4. K-Multivectorial case. Dynkin problem and harmonic extension

We are now in a position to treat the solvability of the problem (III) and problem (IV) stated in the introduction.

Theorem 4.1. *Let $F_k \in C^{0,\nu}(\Gamma)$ be a k-grade multivector field. If $\mathcal{T}_{\Omega_+}\mathcal{D}\tilde{F}_k|_\Gamma$ is k-vector-valued, i.e., if*

$$[\mathcal{T}_{\Omega_+}\mathcal{D}\tilde{F}_k|_\Gamma]_{k\pm 2} = 0, \tag{11}$$

then the problem (III) is solvable.

Proof. From the solution of the jump problem we have that the functions $F^+ := \tilde{F}_k + \mathcal{T}_{\Omega_+}\mathcal{D}\tilde{F}_k$ and $F^- := \mathcal{T}_{\Omega_+}\mathcal{D}\tilde{F}_k$ are monogenic in Ω_+ and Ω_- respectively, and such that on Γ

$$F^+ - F^- = F_k.$$

Moreover, conditions (11) leads to

$$\begin{cases} \Delta[F^\pm]_{k\pm 2} = 0, & \text{in } \Omega_\pm \\ [F^\pm]_{k\pm 2}|_\Gamma = 0. \end{cases}$$

Then, by the classical Dirichlet problem we have $[F^\pm]_{k\pm 2} \equiv 0$ in $\Omega_\pm$, i.e., the functions $F^\pm$ are harmonic k-grade multivector fields in $\Omega_\pm$. □

Theorems 3.1 and 4.1 yield the answer to problem (IV):

Theorem 4.2. *Let $F_k \in C^{0,\nu}(\Gamma)$. If F_k is the trace of a multivector field in $C^{0,\nu}(\Omega_+ \cup \Gamma)$ and harmonic in Ω_+, then*

$$\mathcal{T}_{\Omega_+}\mathcal{D}\tilde{F}_k|_\Gamma = 0 \tag{12}$$

Conversely, if (12) is satisfied, then F_k is the trace of a multivector field in $C^{0,\mu}(\Omega_+ \cup \Gamma)$, $\mu < \nu$, and harmonic in Ω_+.

We would like to mention that all the above theorems in Section 3 and Section 4 can be formally rewritten using surface integration. For that it is enough to employ one of the definitions of integration on a fractal boundary of a Jordan domain, proposed in [20, 21, 24], which is fully discussed in a forthcoming paper [5] by the same authors.

In this more general sense, the Borel-Pompeiu formula allows us to recover a sense for the Cauchy transforms on fractal boundaries, and consequently, all the previously proved sufficient (or necessary) conditions would look like in the smooth context.

Finally we wish to note that, as has been pointed out in [2, 8, 11], there is a closed relation between the harmonic multivector fields and the harmonic differential forms in the sense of Hodge-deRham. Then, the results in Section 4 can be easily adapt to the Hodge-deRahm system theory.

5. Example

Theorem 5.1. *For any* $\mathbf{m} \in [n-1, n)$ *and* $0 < \nu \leq \frac{\mathbf{m}}{n}$ *there exists an hypersurface* Γ_* *such that* $M(\Gamma_*) = \mathbf{m}$ *and a function* $u_* \in C^{0,\nu}(\Gamma_*)$, *such that the jump problem* (3) *has no solution.*

For simplicity we restrict ourselves to surfaces in the 3-dimensional space $\mathbb{R}^3$. In [22] a low-dimensional analogue of the above theorem was proved for complex-valued functions and curves in the plane. Our construction is essentially a higher-dimensional extension of Kats's idea with some necessary modifications.

To prove the theorem we construct a surface $\Gamma_* \subset \mathbb{R}^3$ with box dimension $\mathbf{m}$ which bounds a bounded domain Ω_+ and a function u_* with the following properties:

a) The surface Γ_* contains the origin of coordinates, and any piece of Γ_* not containing this point has finite area.

b) $u_* \in C^{0,\nu}(\overline{\Omega_+})$, where ν is any given number in the interval $(0, \frac{\mathbf{m}}{3}]$, $u_*(x) = 0$ for $x \in \mathbb{R}^n \setminus \overline{\Omega_+}$.

c) $\mathcal{D}u_*(x)$ is bounded outside of any neighborhood of zero, $\mathcal{D}u_*(x) \in L_p$, where $p > 1$.

d) There exist constants $\vartheta < 2$ and $C > 0$ such that

$$|\mathcal{T}_{\Omega_+}\mathcal{D}u_*(x)| \leq C|x|^{-\vartheta}.$$

e) There exist constants $b > 0$, C, such that

$$[\mathcal{T}_{\Omega_+}\mathcal{D}u_*(-x_1e_1)]_1 \geq b\ln(\frac{1}{x_1}) + C,\ 0 < x_1 \leq 1.$$

To prove these properties we shall adapt the arguments from [22]. For the sake of brevity several rather technical steps will be omitted.

Proof of Theorem 5.1. Accept for the moment the validity of these properties and suppose that (3) has a solution Φ for the surface Γ_* and function u_*. We then consider the function

$$U_*(x) = u_*(x) + \mathcal{T}_{\Omega_+}\mathcal{D}u_*(x)$$

and put $\Psi = U_*(x) - \Phi(x)$. By c) $\mathcal{D}U_* \in L_p$, $p > 1$, then

$$\mathcal{D}U_*(x) = \mathcal{D}(u_*(x) + \mathcal{T}_{\Omega_+}\mathcal{D}u_*(x)) = \mathcal{D}u_*(x) - \mathcal{D}u_*(x) = 0$$

and U_* is monogenic in the domains $\Omega_\pm$. On the other hand, since $\mathcal{D}u_*(x)$ is bounded outside of any neighborhood of zero, then it follows that U_* has limiting values $U_*^\pm(z)$ at each point of $\Gamma_* \setminus \{0\}$. Moreover $U_*^+(z) - U_*^-(z) = u(z)$ for $z \in \Gamma_* \setminus \{0\}$ and $U_*(\infty) = 0$. Hence the function Ψ is continuous in $\Gamma_* \setminus \{0\}$ and it is also monogenic in $\mathbb{R}^3 \setminus \Gamma_*$. Then, from a) and the multidimensional Painlevé's Theorem proved in [1] it follows that Ψ is monogenic in $\mathbb{R}^3 \setminus \{0\}$. Next, by d) we have $|\Psi(x)| \leq C|x|^{-\vartheta}$ which implies that the singularity at the point 0 is removable (see [10]). Hence, Ψ is monogenic in $\mathbb{R}^3$ and $\Psi(\infty) = 0$. By Liouville theorem we have $\Psi(x) \equiv 0$ and $U_*(x)$ must be also a solution of (3) which contradicts e). $\square$

5.1. The curve of B. Kats

For the sake of clarity we firstly give a sketch of the Kats's construction, which is borrowed from Reference [22], and leave the details to the reader. The construction is as follows: Fix a number $\beta \geq 2$ (compare with [22]) and denote $M_n = 2^{[n\beta]}$. Suppose $\{a_n^j\}_{j=0}^{M_n}$ are points dividing the interval $I_n = [2^{-n}, 2^{-n+1}]$ into M_n equal parts: $a_n^0 = 2^{-n+1}$, $a_n^1 = 2^{-n+1} - \frac{2^{-n}}{M_n}$, etc. Denote by Λ_n the curve consisting of the vertical intervals $[a_n^j, a_n^j + i2^{-n}]$, $j = 1, \ldots, M_n - 1$, the intervals of the real axis $[a_n^{2j}, a_n^{2j-1}]$, $j = 1, \ldots, \frac{M_n}{2}$, and the horizontal intervals $[a_n^{2j+1} + i2^{-n}, a_n^{2j} + i2^{-n}]$, $j = 1, \ldots, \frac{M_n}{2} - 1$. Then he defined $\Lambda^\beta = \bigcup_{n=1}^\infty \Lambda_n$ and constructed the closed curve γ_0 consisting of the intervals $[0, 1-i]$, $[1-i, 1]$ and the curve Λ^β. Kats showed that $\mathrm{M}(\gamma_0) = \frac{2\beta}{1+\beta}$.

5.2. The surface Γ_*

Denote by σ_0 the closed plane domain bounded by the curve γ_0 lying on the plane $x_3 = 0$ of $\mathbb{R}^3$, and let Γ_0 be the boundary of the three-dimensional closed domain

$$\Omega_0 := \left\{\sigma_0 + \lambda e_3,\ -\frac{1}{2} \leq \lambda \leq \frac{1}{2}\right\}.$$

Next we define the surface Γ_β as the boundary of the closed cut domain

$$\Omega_\beta := \Omega_0 \cap \{2x_3 - x_1 \leq 0\} \cap \{2x_3 + x_1 \geq 0\}$$

As it can be appreciated, the surface Γ_β is composed of four plane pieces belonging to the semi-space $x_2 < 0$ and the upper 'cover' $\mathbf{T} := \Gamma_\beta \cap \{x_2 \geq 0\}$. Denote by $\mathbf{T}_n$ the piece of $\mathbf{T}$ generated by the curve Λ_n. Then we have $\mathbf{T} = \bigcup_{n=1}^\infty \mathbf{T}_n$.

We consider the covering of $\mathbf{T}$ by cubes of the grid $\mathcal{M}_k$. All the sets $\mathbf{T}_{k+1}$, $\mathbf{T}_{k+2}$, ..., are covered by two cubes of this grid. Another two cubes cover $\mathbf{T}_k$. Denote by ρ_n the distance between the vertical intervals of the curve Λ_n, i.e. $\rho_n = \frac{2^{-n}}{M_n} = 2^{-n-[n\beta]}$. To cover the sides of $\mathbf{T}_n$ not more than $2^{3k-3n+1}$ cubes are necessary when $2^{-k} \geq \rho_n$. When $2^{-k} < \rho_n$, to cover the sides which are parallel to the plane $x_1 = 0$

$$\sum_{l=0}^{M_n-1}((2^{-n+1}-\rho_n l)2^{2k-n}) = 2^{2k-1}(3\cdot 2^{[n\beta]-2n}+2^{-n})$$

cubes are necessary; to cover the sides which are parallel to the plane $x_2 = 0$, not more than $2^{2k-2n+2}$ cubes are necessary; and to cover the sides on the planes $2x_3 - x_1 = 0$ and $2x_3 + x_1 = 0$ not more than $2^{2k-2n+1}$ cubes are necessary.

Therefore

$$4 + \sum_{2^{-k}\geq\rho_n,\,k>n} 2^{3k-3n+1} + \sum_{2^{-k}<\rho_n}(2^{2k-1}(3\cdot 2^{[n\beta]-2n}+2^{-2n}))$$
$$+ \sum_{2^{-k}<\rho_n} 2^{2k-2n+2} \geq N_k(\mathbf{T}) \geq \sum_{2^{-k}<\rho_n}(2^{2k-1}(3\cdot 2^{[n\beta]-2n}+2^{-2n}))$$

or more simply

$$4 + \sum_{2^{-k}\geq\rho_n,\,k>n} 2^{-3n} + 6\cdot 2^k \sum_{2^{-k}<\rho_n} 2^{[n\beta]-2n} \geq N_k(\mathbf{T}) \geq \frac{3}{2}\cdot 2^{2k}\sum_{2^{-k}<\rho_n} 2^{[n\beta]-2n}.$$

Denote by B_k the integer defined by the condition

$$\frac{k}{1+\beta} - 1 \leq B_k < \frac{k}{1+\beta}.$$

Suppose that $k > 1+\beta$, then

$$\sum_{2^{-k}<\rho_n} 2^{[n\beta]-2n} = \sum_{n=1}^{B_k} 2^{[n\beta]-2n} \quad \text{and} \quad \sum_{2^{-k}\geq\rho_n,\,k>n} 2^{-3n} = \sum_{n=B_k+1}^{k-1} 2^{-3n}.$$

Simple estimates give

$$C_1\cdot 2^{\frac{3\beta k}{1+\beta}} \leq 2^{2k}\sum_{2^{-k}<\rho_n} 2^{[n\beta]-2n} \leq C_2\cdot 2^{\frac{3\beta k}{1+\beta}}.$$

Analogously

$$2^{3k+1}\sum_{2^{-k}\geq\rho_n,\,k>n} 2^{-3n} \leq C_3\cdot 2^{\frac{3\beta k}{1+\beta}}.$$

From the above estimates we obtain

$$\mathrm{M}(\mathbf{T}) = \frac{3\beta}{1+\beta}.$$

From the simple geometry of the remaining parts of Γ_β we have also

$$\mathrm{M}(\Gamma_\beta) = \frac{3\beta}{1+\beta}.$$

For $2 \leq \mathbf{m} < 3$ we define the desired surface $\Gamma_* := \Gamma_\beta$ with $\beta = \frac{\mathbf{m}}{3-\mathbf{m}}$. Then

$$\mathrm{M}(\Gamma_*) = \mathbf{m}.$$

Finally, we put $\Omega_+ := \Omega_\beta \setminus \Gamma_*$, $\Omega_- := \mathbb{R}^n \setminus \overline{\Omega_\beta}$.

5.3. The function u_*

Let $0 < \nu < 1$. Following the Kats's idea we enumerate all the points a_n^j, $j = 0, \dots, M_n$, $n = 1, 2, \dots$ in decreasing order: $\delta_0 = a_1^0, \delta_1 = a_1^1, \dots$. Denote by $\Delta_k = \delta_k - \delta_{k+1}$. Obviously, $\Delta_0 = \dots = \Delta_{M_1 - 1} = \rho_1$, $\Delta_{M_1} = \dots = \Delta_{M_2-1} = \rho_2$, ..., etc. Since the sequence $\{\Delta_k\}_{k=0}^{\infty}$ is nonincreasing, the series

$$\sum_{j=k}^{\infty} (-1)^j \Delta_j^{\nu}$$

converges. Define the function φ in the following way:

$$\varphi(\delta_k) = \sum_{j=k}^{\infty} (-1)^j \Delta_j^{\nu}.$$

We extend the definition of the function φ to the interval $[0, 1]$ requiring that it be linear on all intervals $[\delta_{k+1}, \delta_k]$, $\varphi(0) = 0$. Further we set

$$\varphi(x) = \varphi(x_1 e_1 + x_2 e_2 + x_3 e_3) = \varphi(x_1)$$

if x lies in the closed domain $\Omega_+ \cup \Gamma_*$, and $\varphi(x) = 0$ for $x \in \Omega_-$. It can be proved (see [22]) that $\varphi \in C^{0,\nu}([0, 1])$, then $\varphi \in C^{0,\nu}(\overline{\Omega_+})$.

Denote

$$\mu = (1 + \beta)(1 - \nu) = \frac{3(1 - \nu)}{3 - \mathbf{m}}, \quad \eta = [\mu] - 1, \quad \frac{3}{2}\vartheta = \mu - \eta.$$

By assumption $\nu \leq \frac{\mathbf{m}}{3}$ we have $\mu \geq 1$,$\eta \geq 0$, and $\frac{2}{3} \leq \vartheta < 2$. Now we define the desired function to be $u_*(x) = x_1^{\eta} \varphi(x)$.

5.4. Proof of properties a) $\cdots$ e)

Property a) is obvious from the construction of Γ_* and the previous calculation of its box dimension. Since $x_1^{\eta} \in C^{0,\nu}(\overline{\Omega +})$ and $\varphi \in C^{0,\nu}(\overline{\Omega_+})$, then $u_* \in C^{0,\nu}(\overline{\Omega_+})$, $u_*(0) = 0$ and furthermore $u_*(x) = 0$ in Ω_-. Hence u_* has property b).

On the other hand

$$\mathcal{D}u_*(x) = e_1 \frac{\partial u_*(x_1)}{\partial x_1} = e_1(\eta x_1^{\eta - 1} \varphi(x_1) + x_1^{\eta} \varphi'(x_1)).$$

For $x_1 \in [2^{-n}, 2^{-n+1}]$ we have $|\varphi'(x_1)| = \rho_n^{\nu - 1} \leq 2^{n\mu}$. Hence $|\varphi'(x_1)| \leq 2^{\mu} x_1^{-\mu}$, $0 < x_1 \leq 1$. Furthermore $\varphi(x_1) \leq \rho_n^{\nu} \leq 2^{\mu} x_1^{1-\mu}$.

From these inequalities we obtain

$$|\mathcal{D}u_*(x)| \leq \eta x_1^{\eta - 1} |\varphi(x_1)| + x_1^{\eta} |\varphi'(x_1)| \leq C x_1^{-\frac{3}{2}\vartheta}.$$

Since for $x \in \overline{\Omega_+}$, $|x| \leq \frac{3}{2}|x_1|$ holds, then

$$|\mathcal{D}u_*(x)| \leq C|x|^{-\frac{3}{2}\vartheta}.$$

Obviously, $\mathcal{D}u_*(x) \in L_p$ for $p \leq \frac{2}{\vartheta}$. Since $\frac{2}{3} \leq \vartheta < 2$ we obtain $\mathcal{D}u_*(x) \in L_p$ for some $p > 1$. Hence property c) follows. The same estimate proves property d). In fact,

$$|x|^\vartheta |\mathcal{T}_{\Omega_+}\mathcal{D}u_*(x)| = \frac{1}{4\pi}|\int_{\Omega_+} \frac{|y-x-y|^\vartheta (x-y)\mathcal{D}u_*(y)}{|y-x|^3} d\mathcal{L}^3(y)|$$

$$\leq \frac{1}{4\pi}|\int_{\Omega_+} \frac{|y-x-y|^\vartheta \mathcal{D}u_*(y)}{|y-x|^2} d\mathcal{L}^3(y)|$$

$$\leq C(\int_{\Omega_+} |y-x|^{\vartheta-2}|\mathcal{D}u_*(y)| d\mathcal{L}^3(y) + \int_{\Omega_+} \frac{|y|^\vartheta |\mathcal{D}u_*(y)|}{|y-x|^2} d\mathcal{L}^3(y)).$$

Using Hölder inequality the two last integrals can be bounded by a constant C independent of x, which proves property d). To prove e) we firstly note that

$$\mathcal{T}_{\Omega_+}(\mathcal{D}u_*)(x) = \eta \mathcal{T}_{\Omega_+}(e_1(\eta x_1^{\eta-1}\varphi(x_1))(x) + \mathcal{T}_{\Omega_+}(e_1 x_1^\eta \varphi'(x_1))(x).$$

Since $\varphi(0) = 0$ and $\varphi \in C^{0,\nu}(\overline{\Omega_+})$, then $x_1^{\eta-1}\varphi(x) \in L_p(\overline{\Omega_+})$ for some $p > 3$ $\mathcal{T}_{\Omega_+}(e_1(\eta x_1^{\eta-1}\varphi(x_1))(x)$ is bounded.

In order to estimate $\mathcal{T}_{\Omega_+}(e_1 x_1^\eta \varphi'(x_1))(x)$ we split it as

$$\mathcal{T}_{\Omega_+}(e_1 x_1^\eta \varphi'(x_1))(x) = \mathcal{T}_{\mathbf{P}}(e_1 x_1^\eta \varphi'(x_1))(x) + \sum_{n=0}^\infty \mathcal{T}_{\Pi_n}(e_1 x_1^\eta \varphi'(x_1))(x)$$

where

$$\mathbf{P} = \{x : \ 0 \leq x_1 \leq 1, \ -x_1 \leq x_2 \leq 0, \ -\frac{1}{2}x_1 \leq x_3 \leq \frac{1}{2}x_1\}$$

and

$$\Pi_n = \{x : \ \delta_{2n+1} \leq x_1 \leq \delta_{2n}, \ 0 \leq x_2 \leq h_{2n}, \ -\frac{1}{2}x_1 \leq x_3 \leq \frac{1}{2}x_1\}$$

$$h_0 = h_1 = \cdots = h_{\frac{M_1}{2}-1} = 2^{-1}, \ h_{\frac{M_1}{2}} = \cdots = h_{\frac{M_2}{2}-1} = 2^{-2}, \ldots.$$

Since P has piecewise smooth boundary ∂P and $u_* \in C^{0,\nu}(\overline{\Omega_+})$, then in virtue of the Borel-Pompeiu formula and the Hölder boundedness of the Cauchy transform $\mathcal{C}_{\partial P}u_*$ (see [1, 7], for instance), we have that $\mathcal{T}_{\mathbf{P}}\mathcal{D}u_*(x)$ is bounded and then $\mathcal{T}_{\mathbf{P}}(e_1 x_1^\eta \varphi'(x_1))(x)$ is also bounded.

The arguments used to state the lower bound for

$$[\sum_{n=0}^\infty \mathcal{T}_{\Pi_n}(e_1 x_1^\eta \varphi'(x_1))(x)]_1$$

are rather technical and follow essentially the same Kats procedure. After that we obtain

$$[\mathcal{T}_{\Omega_+}\mathcal{D}u_*(-x_1 e_1)]_1 \geq C + b\sum_{k=0}^\infty \frac{2^{k(2\vartheta-1)}}{(2^k x_1 + 1)^2}, \ b > 0, \ x_1 \in (0, 1].$$

Since

$$\sum_{k=0}^{\infty} \frac{2^{k(2\vartheta-1)}}{(2^k x_1 + 1)^2} \geq \int_0^{\infty} \frac{dt}{(2^t x_1 + 1)^2} = \int_0^{\infty} \frac{2^{-2t} dt}{(x_1 + 2^{-t})^2} \geq \ln 2 \ln \frac{1}{x_1}$$

then we obtain the desired inequality e).

References

[1] R. Abreu and J. Bory: *Boundary value problems for quaternionic monogenic functions on non-smooth surfaces.* Adv. Appl. Clifford Algebras, **9**, No.1, 1999: 1–22.

[2] R. Abreu, J. Bory, R. Delanghe and F. Sommen: *Harmonic multivector fields and the Cauchy integral decomposition in Clifford analysis.* BMS Simon Stevin, **11**, No 1, 2004: 95–110.

[3] R. Abreu, J. Bory, R. Delanghe and F. Sommen: *Cauchy integral decomposition of multi-vector-valued functions on hypersurfaces.* Computational Methods and Function Theory, Vol. 5, No. 1, 2005: 111–134

[4] R. Abreu, J. Bory, O. Gerus and M. Shapiro: *The Clifford/Cauchy transform with a continuous density; N. Davydov's theorem.* Math. Meth. Appl. Sci, Vol. 28, No. 7, 2005: 811–825.

[5] R. Abreu, J. Bory and T. Moreno: *Integration of multi-vector-valued functions over fractal hypersurfaces.* In preparation.

[6] R. Abreu, D. Peña, J. Bory: *Clifford Cauchy Type Integrals on Ahlfors-David Regular Surfaces in $\mathbb{R}^{m+1}$.* Adv. Appl. Clifford Algebras **13** (2003), No. 2, 133–156.

[7] R. Abreu, J. Bory and D. Peña: *Jump problem and removable singularities for monogenic functions.* Submitted.

[8] R. Abreu, J. Bory and M. Shapiro: *The Cauchy transform for the Hodge/deRham system and some of its properties.* Submitted.

[9] J. Bory and R. Abreu: *Cauchy transform and rectifiability in Clifford Analysis.* Z. Anal. Anwend, **24**, No 1, 2005: 167–178.

[10] F. Brackx, R. Delanghe and F. Sommen: *Clifford analysis.* Research Notes in Mathematics, 76. Pitman (Advanced Publishing Program), Boston, MA, 1982. x+308 pp. ISBN: 0-273-08535-2 MR 85j:30103.

[11] F. Brackx, R. Delanghe and F. Sommen: *Differential Forms and/or Multivector Functions.* Cubo **7** (2005), no. 2, 139–169.

[12] R. Delanghe, F. Sommen and V. Souček: *Clifford algebra and spinor-valued functions. A function theory for the Dirac operator.* Related REDUCE software by F. Brackx and D. Constales. With 1 IBM-PC floppy disk (3.5 inch). Mathematics and its Applications, 53. Kluwer Academic Publishers Group, Dordrecht, 1992. xviii+485 pp. ISBN: 0-7923-0229-X MR 94d:30084.

[13] E. Dyn'kin: *Cauchy integral decomposition for harmonic forms.* Journal d'Analyse Mathématique. 1997; **37**: 165–186.

[14] Falconer, K.J.: *The geometry of fractal sets.* Cambridge Tracts in Mathematics, 85. *Cambridge University Press, Cambridge*, 1986.

[15] Feder, Jens.: *Fractals.* With a foreword by Benoit B. Mandelbrot. Physics of Solids and Liquids. *Plenum Press, New York*, 1988.

[16] H. Federer: *Geometric measure theory.* Die Grundlehren der mathematischen Wissenschaften, Band 153, Springer-Verlag, New York Inc.,(1969) New York, xiv+676 pp.

[17] J. Gilbert, J. Hogan and J. Lakey: *Frame decomposition of form-valued Hardy spaces.* In Clifford Algebras in analysis and related topics, Ed. J. Ryan CRC Press, (1996), Boca Raton, 239–259.

[18] J. Gilbert and M. Murray: *Clifford algebras and Dirac operators in harmonic analysis.* Cambridge Studies in Advanced Mathematics **26**, Cambridge, 1991.

[19] K. Gürlebeck and W. Sprössig: *Quaternionic and Clifford Calculus for Physicists and Engineers.* Wiley and Sons Publ., 1997.

[20] J. Harrison and A. Norton: *Geometric integration on fractal curves in the plane.* Indiana University Mathematical Journal, **4**, No. 2, 1992: 567–594.

[21] J. Harrison and A. Norton: *The Gauss-Green theorem for fractal boundaries.* Duke Mathematical Journal, **67**, No. 3, 1992: 575–588.

[22] B.A. Kats: *The Riemann problem on a closed Jordan curve.* Izv. Vuz. Matematika, **27**, No. 4, 1983: 68–80.

[23] B.A. Kats: *On the solvability of the Riemann boundary value problem on a fractal arc.* (Russian) *Mat. Zametki* 53 (1993), no. 5, 69–75; *translation in Math. Notes* 53 (1993), No. 5-6, 502–505.

[24] B.A. Kats: *Jump problem and integral over nonrectifiable curve.* Izv. Vuz. Matematika **31**, No. 5, 1987: 49–57.

[25] M.L. Lapidus and H. Maier: *Hypothèse de Riemann, cordes fractales vibrantes et conjecture de Weyl-Berry modifiée. (French) [The Riemann hypothesis, vibrating fractal strings and the modified Weyl-Berry conjecture] C.R. Acad. Sci. Paris Sér. I Math. 313 (1991), no. 1, 19–24.* 1991.

[26] E.M. Stein: *Singular integrals and differentiability properties of functions.* Princeton Math. Ser. **30**, Princeton Univ. Press, Princeton, N.J. 1970.

Ricardo Abreu-Blaya
Facultad de Informatica y Matematica
Universidad de Holguin, Cuba
e-mail: `rabreu@facinf.uho.edu.cu`

Juan Bory-Reyes
Departamento of Matematica
Universidad de Oriente, Cuba
e-mail: `jbory@rect.uo.edu.cu`

Tania Moreno-García
Facultad de Informatica y Matematica
Universidad de Holguin, Cuba
e-mail: `tania.moreno@facinf.uho.edu.cu`

Operator Theory:
Advances and Applications, Vol. 167, 17–67

Metric Dependent Clifford Analysis with Applications to Wavelet Analysis

Fred Brackx, Nele De Schepper and Frank Sommen

Abstract. In earlier research multi-dimensional wavelets have been constructed in the framework of Clifford analysis. Clifford analysis, centered around the notion of monogenic functions, may be regarded as a direct and elegant generalization to higher dimension of the theory of the holomorphic functions in the complex plane. This Clifford wavelet theory might be characterized as isotropic, since the metric in the underlying space is the standard Euclidean one.

In this paper we develop the idea of a metric dependent Clifford analysis leading to a so-called anisotropic Clifford wavelet theory featuring wavelet functions which are adaptable to preferential, not necessarily orthogonal, directions in the signals or textures to be analyzed.

Mathematics Subject Classification (2000). 42B10; 44A15; 30G35.

Keywords. Continuous Wavelet Transform; Clifford analysis; Hermite polynomials.

1. Introduction

During the last fifty years, Clifford analysis has gradually developed to a comprehensive theory which offers a direct, elegant and powerful generalization to higher dimension of the theory of holomorphic functions in the complex plane. Clifford analysis focuses on so-called *monogenic functions*, which are in the simple but useful setting of flat m-dimensional Euclidean space, null solutions of the Clifford-vector valued Dirac operator

$$\underline{\partial} = \sum_{j=1}^{m} e_j \partial_{x_j},$$

where $(e_1, \ldots, e_m)$ forms an orthogonal basis for the quadratic space $\mathbb{R}^m$ underlying the construction of the real Clifford algebra $\mathbb{R}_m$. Numerous papers, conference proceedings and books have moulded this theory and shown its ability for applications, let us mention [2, 15, 18, 19, 20, 24, 25, 26].

Clifford analysis is closely related to harmonic analysis in that monogenic functions refine the properties of harmonic functions. Note for instance that each harmonic function $h(\underline{x})$ can be split as $h(\underline{x}) = f(\underline{x}) + \underline{x}\ g(\underline{x})$ with f, g monogenic, and that a real harmonic function is always the real part of a monogenic one, which does not need to be the case for a harmonic function of several complex variables. The reason for this intimate relationship is that, as does the Cauchy-Riemann operator in the complex plane, the rotation-invariant Dirac operator factorizes the m-dimensional Laplace operator.

A highly important intrinsic feature of Clifford analysis is that it encompasses all dimensions at once, in other words all concepts in this multi-dimensional theory are not merely tensor products of one dimensional phenomena but are directly defined and studied in multi-dimensional space and cannot be recursively reduced to lower dimension. This true multi-dimensional nature has allowed for among others a very specific and original approach to multi-dimensional wavelet theory.

Wavelet analysis is a particular time- or space-scale representation of functions, which has found numerous applications in mathematics, physics and engineering (see, e.g., [12, 14, 21]). Two of the main themes in wavelet theory are the Continuous Wavelet Transform (abbreviated CWT) and discrete orthonormal wavelets generated by multiresolution analysis. They enjoy more or less opposite properties and both have their specific field of application. The CWT plays an analogous rôle as the Fourier transform and is a successful tool for the analysis of signals and feature detection in signals. The discrete wavelet transform is the analogue of the Discrete Fourier Transform and provides a powerful technique for, e.g., data compression and signal reconstruction. It is only the CWT we are aiming at as an application of the theory developed in this paper.

Let us first explain the idea of the one dimensional CWT. Wavelets constitute a family of functions $\psi_{a,b}$ derived from one single function ψ, called the mother wavelet, by change of scale a (i.e., by dilation) and by change of position b (i.e., by translation):

$$\psi_{a,b}(x) = \frac{1}{\sqrt{a}}\, \psi\left(\frac{x-b}{a}\right), \quad a > 0, \quad b \in \mathbb{R}.$$

In wavelet theory some conditions on the mother wavelet ψ have to be imposed. We request ψ to be an L_2-function (finite energy signal) which is well localized both in the time domain and in the frequency domain. Moreover it has to satisfy the so-called admissibility condition:

$$C_\psi := \int_{-\infty}^{+\infty} \frac{|\widehat{\psi}(u)|^2}{|u|}\, du < +\infty,$$

where $\widehat{\psi}$ denotes the Fourier transform of ψ. In the case where ψ is also in L_1, this admissibility condition implies

$$\int_{-\infty}^{+\infty} \psi(x)\, dx = 0.$$

In other words: ψ must be an oscillating function, which explains its qualification as "wavelet".

In practice, applications impose additional requirements, among which a given number of vanishing moments:

$$\int_{-\infty}^{+\infty} x^n \, \psi(x) \, dx = 0, \quad n = 0, 1, \ldots, N.$$

This means that the corresponding CWT:

$$\begin{aligned} F(a,b) &= \langle \psi_{a,b}, f \rangle \\ &= \frac{1}{\sqrt{a}} \int_{-\infty}^{+\infty} \overline{\psi} \left(\frac{x-b}{a} \right) f(x) \, dx \end{aligned}$$

will filter out polynomial behavior of the signal f up to degree N, making it adequate at detecting singularities.

The CWT may be extended to higher dimension while still enjoying the same properties as in the one-dimensional case. Traditionally these higher-dimensional CWTs originate as tensor products of one-dimensional phenomena. However also the non-separable treatment of two-dimensional wavelets should be mentioned (see [1]).

In a series of papers [4, 5, 6, 7, 9, 10] multi-dimensional wavelets have been constructed in the framework of Clifford analysis. These wavelets are based on Clifford generalizations of the Hermite polynomials, the Gegenbauer polynomials, the Laguerre polynomials and the Jacobi polynomials. Moreover, they arise as specific applications of a general theory for constructing multi-dimensional Clifford-wavelets (see [8]). The first step in this construction method is the introduction of new polynomials, generalizing classical orthogonal polynomials on the real line to the Clifford analysis setting. Their construction rests upon a specific Clifford analysis technique, the so-called Cauchy-Kowalewskaia extension of a real-analytic function in $\mathbb{R}^m$ to a monogenic function in $\mathbb{R}^{m+1}$. One starts from a real-analytic function in an open connected domain in $\mathbb{R}^m$, as an analogon of the classical weight function. The new Clifford algebra-valued polynomials are then generated by the Cauchy-Kowalewskaia extension of this weight function. For these polynomials a recurrence relation and a Rodrigues formula are established. This Rodrigues formula together with Stokes's theorem lead to an orthogonality relation of the new Clifford-polynomials. From this orthogonality relation we select candidates for mother wavelets and show that these candidates indeed may serve as kernel functions for a multi-dimensional CWT if they satisfy certain additional conditions.

The above sketched Clifford wavelet theory may be characterized as isotropic since the metric in the underlying space is the standard Euclidean one for which

$$e_j^2 = -1, \quad j = 1, \ldots, m$$

and

$$e_j e_k = -e_k e_j, \quad 1 \leq j \neq k \leq m,$$

leading to the standard scalar product of a vector $\underline{x} = \sum_{j=1}^{m} x_j e_j$ with itself:

$$\langle \underline{x}, \underline{x} \rangle = \sum_{j=1}^{m} x_j^2. \tag{1.1}$$

In this paper we develop the idea of a metric dependent Clifford analysis leading to a so-called anisotropic Clifford wavelet theory featuring wavelet functions which are adaptable to preferential, not necessarily orthogonal, directions in the signals or textures to be analyzed. This is achieved by considering functions taking their values in a Clifford algebra which is constructed over $\mathbb{R}^m$ by means of a symmetric bilinear form such that the scalar product of a vector with itself now takes the form

$$\langle \underline{x}, \underline{x} \rangle = \sum_{j=1}^{m} \sum_{k=1}^{m} g_{jk} x^j x^k. \tag{1.2}$$

We refer to the tensor g_{jk} as the *metric tensor* of the Clifford algebra considered, and it is assumed that this metric tensor is real, symmetric and positive definite. This idea is in fact not completely new since Clifford analysis on manifolds with local metric tensors was already considered in, e.g., [13], [18] and [23], while in [17] a specific three dimensional tensor leaving the third dimension unaltered was introduced for analyzing two dimensional signals and textures. What is new is the detailed development of this Clifford analysis in a global metric dependent setting, the construction of new Clifford-Hermite polynomials and the study of the corresponding Continuous Wavelet Transform. It should be clear that this paper opens a new area in Clifford analysis offering a framework for a new kind of applications, in particular concerning anisotropic Clifford wavelets. We have in mind constructing specific wavelets for analyzing multi-dimensional textures or signals which show more or less constant features in preferential directions. For that purpose we will have the orientation of the fundamental $(e_1, \ldots, e_m)$-frame adapted to these directions resulting in an associated metric tensor which will leave these directions unaltered.

The outline of the paper is as follows. We start with constructing, by means of Grassmann generators, two bases: a covariant one $(e_j \ : \ j = 1, \ldots, m)$ and a contravariant one $(e^j \ : \ j = 1, \ldots, m)$, satisfying the general Clifford algebra multiplication rules:

$$e_j e_k + e_k e_j = -2 g_{jk} \quad \text{and} \quad e^j e^k + e^k e^j = -2 g^{jk}, \quad 1 \leq j, k \leq m$$

with g_{jk} the metric tensor. The above multiplication rules lead in a natural way to the substitution for the classical scalar product (1.1) of a vector $\underline{x} = \sum_{j=1}^{m} x^j e_j$ with itself, a symmetric bilinear form expressed by (1.2). Next we generalize all necessary definitions and results of orthogonal Clifford analysis to this metric dependent setting. In this new context we introduce for, e.g., the concepts of Fischer inner product, Fischer duality, monogenicity and spherical monogenics. Similar to the orthogonal case, we can also in the metric dependent Clifford analysis decompose each homogeneous polynomial into spherical monogenics, which is referred to as the *monogenic decomposition*. After a thorough investigation in the metric dependent context of the so-called Euler and angular Dirac operators, which constitute two fundamental operators in Clifford analysis, we proceed with the introduction of the notions of harmonicity and spherical harmonics. Furthermore, we verify the orthogonal decomposition of homogeneous polynomials into harmonic ones, the so-called *harmonic decomposition*. We end Section 2 with the definition and study of the so-called g-Fourier transform, the metric dependent analogue of the classical Fourier transform.

With a view to integration on hypersurfaces in the metric dependent setting, we invoke the theory of differential forms. In Subsection 3.1 we gather some basic definitions and properties concerning Clifford differential forms. We discuss, e.g., fundamental operators such as the exterior derivative and the basic contraction operators; we also state Stokes's theorem, a really fundamental result in mathematical analysis. Special attention is paid to the properties of the Leray and sigma differential forms, since they both play a crucial rôle in establishing orthogonality relations between spherical monogenics on the unit sphere, the topic of Subsection 3.2.

In a fourth and final section we construct the so-called radial g-Clifford-Hermite polynomials, the metric dependent analogue of the radial Clifford-Hermite polynomials of orthogonal Clifford analysis. For these polynomials a recurrence and orthogonality relation are established. Furthermore, these polynomials turn out to be the desired building blocks for specific wavelet kernel functions, the so-called g-Clifford-Hermite wavelets. We end the paper with the introduction of the corresponding g-Clifford-Hermite Continuous Wavelet Transform.

2. The metric dependent Clifford toolbox

2.1. Tensors

Let us start by recalling a few concepts concerning tensors. Assume in Euclidean space two coordinate systems $(x^1, x^2, \ldots, x^N)$ and $(\widetilde{x}^1, \widetilde{x}^2, \ldots, \widetilde{x}^N)$ given.

Definition 2.1. If $(A^1, A^2, \ldots, A^N)$ in coordinate system $(x^1, x^2, \ldots, x^N)$ and $(\widetilde{A}^1, \widetilde{A}^2, \ldots, \widetilde{A}^N)$ in coordinate system $(\widetilde{x}^1, \widetilde{x}^2, \ldots, \widetilde{x}^N)$ are related by the transformation equations:

$$\widetilde{A}^j = \sum_{k=1}^{N} \frac{\partial \widetilde{x}^j}{\partial x^k} A^k \quad , \quad j = 1, \ldots, N,$$

then they are said to be components of a *contravariant vector*.

Definition 2.2. If $(A_1, A_2, \ldots, A_N)$ in coordinate system $(x^1, x^2, \ldots, x^N)$ and $(\widetilde{A}_1, \widetilde{A}_2, \ldots, \widetilde{A}_N)$ in coordinate system $(\widetilde{x}^1, \widetilde{x}^2, \ldots, \widetilde{x}^N)$ are related by the transformation equations:

$$\widetilde{A}_j = \sum_{k=1}^{N} \frac{\partial x^k}{\partial \widetilde{x}^j} A_k, \quad j = 1, \ldots, N,$$

then they are said to be components of a *covariant vector.*

Example. The sets of differentials $\{dx^1, \ldots, dx^N\}$ and $\{d\widetilde{x}^1, \ldots, d\widetilde{x}^N\}$ transform according to the chain rule:

$$d\widetilde{x}^j = \sum_{k=1}^{N} \frac{\partial \widetilde{x}^j}{\partial x^k} dx^k, \quad j = 1, \ldots, N.$$

Hence $(dx^1, \ldots, dx^N)$ is a contravariant vector.

Example. Consider the coordinate transformation

$$\left(\widetilde{x}^1, \widetilde{x}^2, \ldots, \widetilde{x}^N\right) = \left(x^1, x^2, \ldots, x^N\right) \ A$$

with $A = (a_k^j)$ an $(N \times N)$-matrix. We have

$$\widetilde{x}^j = \sum_{k=1}^{N} x^k a_k^j \quad \text{or equivalently} \quad \widetilde{x}^j = \sum_{k=1}^{N} \frac{\partial \widetilde{x}^j}{\partial x^k} x^k,$$

which implies that $(x^1, \ldots, x^N)$ is a contravariant vector.

Definition 2.3. The *outer tensorial product* of two vectors is a *tensor of rank 2.* There are three possibilities:

- the outer product of two contravariant vectors $(A^1,\ldots,A^N)$ and $(B^1,\ldots,B^N)$ is a contravariant tensor of rank 2:

$$C^{jk} = A^j B^k$$

- the outer product of a covariant vector $(A_1, \ldots, A_N)$ and a contravariant vector $(B^1, \ldots, B^N)$ is a mixed tensor of rank 2:

$$C_j^k = A_j B^k$$

- the outer product of two covariant vectors $(A_1, \ldots, A_N)$ and $(B_1, \ldots, B_N)$ is a covariant tensor of rank 2:

$$C_{jk} = A_j B_k.$$

Example. The Kronecker-delta

$$\delta_j^k = \begin{cases} 1 & \text{if } j = k \\ 0 & \text{otherwise} \end{cases}$$

is a mixed tensor of rank 2.

Example. The transformation matrix

$$a_k^j = \frac{\partial \widetilde{x}^j}{\partial x^k}, \quad j, k = 1, \dots, N$$

is also a mixed tensor of rank 2.

Remark 2.4. In view of Definitions 2.1 and 2.2 it is easily seen that the transformation formulae for tensors of rank 2 take the following form:

$$\begin{aligned}
\widetilde{C}^{jk} = \widetilde{A}^j \widetilde{B}^k &= \sum_{i=1}^{N} \frac{\partial \widetilde{x}^j}{\partial x^i} A^i \sum_{\ell=1}^{N} \frac{\partial \widetilde{x}^k}{\partial x^\ell} B^\ell \\
&= \sum_{i=1}^{N} \sum_{\ell=1}^{N} \frac{\partial \widetilde{x}^j}{\partial x^i} \frac{\partial \widetilde{x}^k}{\partial x^\ell} A^i B^\ell = \sum_{i=1}^{N} \sum_{\ell=1}^{N} \frac{\partial \widetilde{x}^j}{\partial x^i} \frac{\partial \widetilde{x}^k}{\partial x^\ell} C^{i\ell}
\end{aligned}$$

and similarly

$$\widetilde{C}_j^k = \sum_{i=1}^{N} \sum_{\ell=1}^{N} \frac{\partial x^i}{\partial \widetilde{x}^j} \frac{\partial \widetilde{x}^k}{\partial x^\ell} C_i^\ell, \qquad \widetilde{C}_{jk} = \sum_{i=1}^{N} \sum_{\ell=1}^{N} \frac{\partial x^i}{\partial \widetilde{x}^j} \frac{\partial x^\ell}{\partial \widetilde{x}^k} C_{i\ell}.$$

Definition 2.5. The *tensorial contraction* of a tensor of rank p is a tensor of rank $(p-2)$ which one obtains by summation over a common contravariant and covariant index.

Example. The tensorial contraction of the mixed tensor C_k^j of rank 2 is a tensor of rank 0, i.e., a scalar:

$$\sum_{j=1}^{N} C_j^j = D,$$

while the tensorial contraction of the mixed tensor C_i^{jk} of rank 3 yields a contravariant vector:

$$\sum_{j=1}^{N} C_j^{jk} = D^k.$$

Definition 2.6. The *inner tensorial product* of two vectors is their outer product followed by contraction.

Example. The inner tensorial product of the covariant vector $(A_1, \dots, A_N)$ and the contravariant vector $(B^1, \dots, B^N)$ is the tensor of rank 0, i.e., the scalar given by:

$$\sum_{j=1}^{N} A_j B^j = C.$$

2.2. From Grassmann to Clifford

We consider the Grassmann algebra Λ generated by the basis elements $(\mathfrak{f}_j,\ j = 1,\ldots,m)$ satisfying the relations

$$\mathfrak{f}_j\mathfrak{f}_k + \mathfrak{f}_k\mathfrak{f}_j = 0 \quad 1 \leq j,k \leq m.$$

These basis elements form a covariant tensor $(\mathfrak{f}_1,\mathfrak{f}_2,\ldots,\mathfrak{f}_m)$ of rank 1.

Next we consider the dual Grassmann algebra Λ^+ generated by the dual basis $(\mathfrak{f}^{+j}, j = 1,\ldots,m)$, forming a contravariant tensor of rank 1 and satisfying the Grassmann identities:

$$\mathfrak{f}^{+j}\mathfrak{f}^{+k} + \mathfrak{f}^{+k}\mathfrak{f}^{+j} = 0 \quad 1 \leq j,k \leq m. \tag{2.1}$$

Duality between both Grassmann algebras is expressed by:

$$\mathfrak{f}_j\mathfrak{f}^{+k} + \mathfrak{f}^{+k}\mathfrak{f}_j = \delta_j^k. \tag{2.2}$$

Note that both the left- and right-hand side of the above equation is a mixed tensor of rank 2.

Now we introduce the fundamental covariant tensor g_{jk} of rank 2. It is assumed to have real entries, to be positive definite and symmetric:

$$g_{jk} = g_{kj}, \quad 1 \leq j,k \leq m.$$

Definition 2.7. The real, positive definite and symmetric tensor g_{jk} is called the *metric tensor.*

Its reciprocal tensor (a contravariant one) is given by

$$g^{jk} = \frac{1}{\det(g_{jk})} G^{jk},$$

where G^{jk} denotes the cofactor of g_{jk}. It thus satisfies

$$\sum_{\ell=1}^{m} g^{j\ell} g_{\ell k} = \delta_k^j.$$

In what follows we will use the Einstein summation convention, i.e., summation over equal contravariant and covariant indices is tacitly understood.

With this convention the above equation expressing reciprocity is written as

$$g^{j\ell} g_{\ell k} = \delta_k^j.$$

Definition 2.8. The covariant basis $(\mathfrak{f}_j^+, j = 1,\ldots,m)$ for the Grassmann algebra Λ^+ is given by

$$\mathfrak{f}_j^+ = g_{jk}\mathfrak{f}^{+k}.$$

This covariant basis shows the following properties.

Proposition 2.9. *One has*

$$\mathfrak{f}_j^+\mathfrak{f}_k^+ + \mathfrak{f}_k^+\mathfrak{f}_j^+ = 0 \quad \text{and} \quad \mathfrak{f}_j\mathfrak{f}_k^+ + \mathfrak{f}_k^+\mathfrak{f}_j = g_{jk}\ ; \quad 1 \leq j,k \leq m.$$

Proof. By means of respectively (2.1) and (2.2), we find

$$\begin{aligned}\mathfrak{f}_j^+\mathfrak{f}_k^+ + \mathfrak{f}_k^+\mathfrak{f}_j^+ &= g_{j\ell}\mathfrak{f}^{+\ell}g_{kt}\mathfrak{f}^{+t} + g_{kt}\mathfrak{f}^{+t}g_{j\ell}\mathfrak{f}^{+\ell}\\ &= g_{j\ell}g_{kt}(\mathfrak{f}^{+\ell}\mathfrak{f}^{+t} + \mathfrak{f}^{+t}\mathfrak{f}^{+\ell}) = 0\end{aligned}$$

and

$$\begin{aligned}\mathfrak{f}_j\mathfrak{f}_k^+ + \mathfrak{f}_k^+\mathfrak{f}_j &= \mathfrak{f}_j g_{kt}\mathfrak{f}^{+t} + g_{kt}\mathfrak{f}^{+t}\mathfrak{f}_j = g_{kt}(\mathfrak{f}_j\mathfrak{f}^{+t} + \mathfrak{f}^{+t}\mathfrak{f}_j)\\ &= g_{kt}\delta_j^t = g_{kj} = g_{jk}.\end{aligned}$$

□

Remark 2.10. By reciprocity one has:

$$\mathfrak{f}^{+k} = g^{kj}\mathfrak{f}_j^+.$$

Definition 2.11. The contravariant basis ($\mathfrak{f}^j$, $j = 1,\ldots,m$) for the Grassmann algebra Λ is given by

$$\mathfrak{f}^j = g^{jk}\mathfrak{f}_k.$$

It shows the following properties.

Proposition 2.12. *One has*

$$\mathfrak{f}^j\mathfrak{f}^k + \mathfrak{f}^k\mathfrak{f}^j = 0 \quad \text{and} \quad \mathfrak{f}^j\mathfrak{f}^{+k} + \mathfrak{f}^{+k}\mathfrak{f}^j = g^{jk}.$$

Proof. A straightforward computation yields

$$\begin{aligned}\mathfrak{f}^j\mathfrak{f}^k + \mathfrak{f}^k\mathfrak{f}^j &= g^{j\ell}\mathfrak{f}_\ell g^{kt}\mathfrak{f}_t + g^{kt}\mathfrak{f}_t g^{j\ell}\mathfrak{f}_\ell\\ &= g^{j\ell}g^{kt}(\mathfrak{f}_\ell\mathfrak{f}_t + \mathfrak{f}_t\mathfrak{f}_\ell) = 0\end{aligned}$$

and

$$\begin{aligned}\mathfrak{f}^j\mathfrak{f}^{+k} + \mathfrak{f}^{+k}\mathfrak{f}^j &= g^{j\ell}\mathfrak{f}_\ell\mathfrak{f}^{+k} + \mathfrak{f}^{+k}g^{j\ell}\mathfrak{f}_\ell\\ &= g^{j\ell}(\mathfrak{f}_\ell\mathfrak{f}^{+k} + \mathfrak{f}^{+k}\mathfrak{f}_\ell)\\ &= g^{j\ell}\delta_\ell^k = g^{jk}.\end{aligned}$$

□

Remark 2.13. By reciprocity one has:

$$\mathfrak{f}_k = g_{kj}\mathfrak{f}^j.$$

Now we consider in the direct sum

$$\mathrm{span}_{\mathbb{C}}\{\mathfrak{f}_1,\ldots,\mathfrak{f}_m\} \oplus \mathrm{span}_{\mathbb{C}}\{\mathfrak{f}_1^+,\ldots,\mathfrak{f}_m^+\}$$

two subspaces, viz. $\mathrm{span}_{\mathbb{C}}\{e_1,\ldots,e_m\}$ and $\mathrm{span}_{\mathbb{C}}\{e_{m+1},\ldots,e_{2m}\}$ where the new covariant basis elements are defined by

$$\begin{cases} e_j &= \mathfrak{f}_j - \mathfrak{f}_j^+, \quad j = 1,\ldots,m\\ e_{m+j} &= i(\mathfrak{f}_j + \mathfrak{f}_j^+), \quad j = 1,\ldots,m.\end{cases}$$

Similarly, we consider the contravariant reciprocal subspaces $\mathrm{span}_{\mathbb{C}}\{e^1,\ldots,e^m\}$ and $\mathrm{span}_{\mathbb{C}}\{e^{m+1},\ldots,e^{2m}\}$ given by

$$\begin{cases} e^j &= \mathfrak{f}^j - \mathfrak{f}^{+j}, \quad j = 1,\ldots,m\\ e^{m+j} &= i(\mathfrak{f}^j + \mathfrak{f}^{+j}), \quad j = 1,\ldots,m.\end{cases}$$

The covariant basis shows the following properties.

Proposition 2.14. *For $1 \leq j, k \leq m$ one has:*

(i) $e_j e_k + e_k e_j = -2g_{jk}$
(ii) $e_{m+j} e_{m+k} + e_{m+k} e_{m+j} = -2g_{jk}$
(iii) $e_j e_{m+k} + e_{m+k} e_j = 0.$

Proof. A straightforward computation leads to

(i)
$$\begin{aligned} e_j e_k + e_k e_j &= (\mathfrak{f}_j - \mathfrak{f}_j^+)(\mathfrak{f}_k - \mathfrak{f}_k^+) + (\mathfrak{f}_k - \mathfrak{f}_k^+)(\mathfrak{f}_j - \mathfrak{f}_j^+) \\ &= \mathfrak{f}_j\mathfrak{f}_k - \mathfrak{f}_j\mathfrak{f}_k^+ - \mathfrak{f}_j^+\mathfrak{f}_k + \mathfrak{f}_j^+\mathfrak{f}_k^+ + \mathfrak{f}_k\mathfrak{f}_j - \mathfrak{f}_k\mathfrak{f}_j^+ - \mathfrak{f}_k^+\mathfrak{f}_j + \mathfrak{f}_k^+\mathfrak{f}_j^+ \\ &= (\mathfrak{f}_j\mathfrak{f}_k + \mathfrak{f}_k\mathfrak{f}_j) + (\mathfrak{f}_j^+\mathfrak{f}_k^+ + \mathfrak{f}_k^+\mathfrak{f}_j^+) - (\mathfrak{f}_j\mathfrak{f}_k^+ + \mathfrak{f}_k^+\mathfrak{f}_j) - (\mathfrak{f}_j^+\mathfrak{f}_k + \mathfrak{f}_k\mathfrak{f}_j^+) \\ &= -g_{jk} - g_{kj} = -2g_{jk}, \end{aligned}$$

(ii)
$$\begin{aligned} & e_{m+j}e_{m+k} + e_{m+k}e_{m+j} \\ &= -(\mathfrak{f}_j + \mathfrak{f}_j^+)(\mathfrak{f}_k + \mathfrak{f}_k^+) - (\mathfrak{f}_k + \mathfrak{f}_k^+)(\mathfrak{f}_j + \mathfrak{f}_j^+) \\ &= -\mathfrak{f}_j\mathfrak{f}_k - \mathfrak{f}_j\mathfrak{f}_k^+ - \mathfrak{f}_j^+\mathfrak{f}_k - \mathfrak{f}_j^+\mathfrak{f}_k^+ - \mathfrak{f}_k\mathfrak{f}_j - \mathfrak{f}_k\mathfrak{f}_j^+ - \mathfrak{f}_k^+\mathfrak{f}_j - \mathfrak{f}_k^+\mathfrak{f}_j^+ \\ &= -(\mathfrak{f}_j\mathfrak{f}_k^+ + \mathfrak{f}_k^+\mathfrak{f}_j) - (\mathfrak{f}_j^+\mathfrak{f}_k + \mathfrak{f}_k\mathfrak{f}_j^+) \\ &= -g_{jk} - g_{kj} = -2g_{jk} \end{aligned}$$

and

(iii)
$$\begin{aligned} & e_j e_{m+k} + e_{m+k} e_j \\ &= i(\mathfrak{f}_j - \mathfrak{f}_j^+)(\mathfrak{f}_k + \mathfrak{f}_k^+) + i(\mathfrak{f}_k + \mathfrak{f}_k^+)(\mathfrak{f}_j - \mathfrak{f}_j^+) \\ &= i\mathfrak{f}_j\mathfrak{f}_k + i\mathfrak{f}_j\mathfrak{f}_k^+ - i\mathfrak{f}_j^+\mathfrak{f}_k - i\mathfrak{f}_j^+\mathfrak{f}_k^+ + i\mathfrak{f}_k\mathfrak{f}_j - i\mathfrak{f}_k\mathfrak{f}_j^+ + i\mathfrak{f}_k^+\mathfrak{f}_j - i\mathfrak{f}_k^+\mathfrak{f}_j^+ \\ &= i(\mathfrak{f}_j\mathfrak{f}_k + \mathfrak{f}_k\mathfrak{f}_j) - i(\mathfrak{f}_j^+\mathfrak{f}_k^+ + \mathfrak{f}_k^+\mathfrak{f}_j^+) + i(\mathfrak{f}_j\mathfrak{f}_k^+ + \mathfrak{f}_k^+\mathfrak{f}_j) - i(\mathfrak{f}_j^+\mathfrak{f}_k + \mathfrak{f}_k\mathfrak{f}_j^+) \\ &= ig_{jk} - ig_{kj} = 0. \end{aligned}$$

□

As expected, both e-bases are linked to each other by means of the metric tensor g_{jk} and its reciprocal g^{jk}.

Proposition 2.15. *For $j = 1, \ldots, m$ one has*

(i)
$$e^j = g^{jk} e_k \quad \text{and} \quad e_k = g_{kj} e^j$$

(ii)
$$e^{m+j} = g^{jk} e_{m+k} \quad \text{and} \quad e_{m+k} = g_{kj} e^{m+j}.$$

Proof.

(i)
$$\begin{aligned} e^j &= \mathfrak{f}^j - \mathfrak{f}^{+j} = g^{jk}\mathfrak{f}_k - g^{jk}\mathfrak{f}_k^+ \\ &= g^{jk}(\mathfrak{f}_k - \mathfrak{f}_k^+) = g^{jk} e_k \end{aligned}$$

(ii)
$$\begin{aligned} e^{m+j} &= i(\mathfrak{f}^j + \mathfrak{f}^{+j}) = ig^{jk}(\mathfrak{f}_k + \mathfrak{f}_k^+) \\ &= g^{jk} e_{m+k}. \end{aligned}$$

□

By combining Propositions 2.14 and 2.15 we obtain the following properties of the contravariant e-basis.

Proposition 2.16. *For $1 \leq j, k \leq m$ one has*
(i) $e^j e^k + e^k e^j = -2g^{jk}$
(ii) $e^{m+j} e^{m+k} + e^{m+k} e^{m+j} = -2g^{jk}$
(iii) $e^j e^{m+k} + e^{m+k} e^j = 0.$

The basis $(e_1, \ldots, e_m)$ and the dual basis $(e^1, \ldots, e^m)$ are also linked to each other by the following relations.

Proposition 2.17. *For $1 \leq j, k \leq m$ one has*
(i) $e_j e^k + e^k e_j = -2\delta_j^k$
(ii) $e_{m+j} e^{m+k} + e^{m+k} e_{m+j} = -2\delta_j^k$
(iii) $e_j e^{m+k} + e^{m+k} e_j = 0$
(iv) $\sum_j e_j e^j = \sum_j e^j e_j = -m$
(v) $\frac{1}{2} \sum_j \{e_j, e^j\} = -m$
(vi) $\sum_j [e_j, e^j] = 0$
(vii) $\sum_j e_{m+j} e^{m+j} = \sum_j e^{m+j} e_{m+j} = -m$
(viii) $\frac{1}{2} \sum_j \{e_{m+j}, e^{m+j}\} = -m$
(ix) $\sum_j [e_{m+j}, e^{m+j}] = 0.$

Proof. A straightforward computation yields

(i)
$$\begin{aligned} e_j e^k + e^k e_j &= e_j g^{kt} e_t + g^{kt} e_t e_j = g^{kt}(e_j e_t + e_t e_j) \\ &= g^{kt}(-2g_{jt}) = -2\delta_j^k \end{aligned}$$

(ii)
$$\begin{aligned} e_{m+j} e^{m+k} + e^{m+k} e_{m+j} &= e_{m+j} g^{kt} e_{m+t} + g^{kt} e_{m+t} e_{m+j} \\ &= g^{kt}(e_{m+j} e_{m+t} + e_{m+t} e_{m+j}) \\ &= g^{kt}(-2g_{jt}) = -2\delta_j^k \end{aligned}$$

(iii)
$$\begin{aligned} e_j e^{m+k} + e^{m+k} e_j &= e_j g^{kt} e_{m+t} + g^{kt} e_{m+t} e_j \\ &= g^{kt}(e_j e_{m+t} + e_{m+t} e_j) = 0 \end{aligned}$$

(iv)
$$\begin{aligned} \sum_{j=1}^m e_j e^j &= \sum_{j=1}^m e_j \sum_{i=1}^m g^{ji} e_i = \sum_{i,j} g^{ij} e_j e_i \\ &= \frac{1}{2} \sum_{i,j} g^{ij} e_j e_i + \frac{1}{2} \sum_{i,j} g^{ij} e_j e_i = \frac{1}{2} \sum_{i,j} g^{ij} e_j e_i + \frac{1}{2} \sum_{j,i} g^{ji} e_i e_j \\ &= \frac{1}{2} \sum_{i,j} g^{ij} e_j e_i + \frac{1}{2} \sum_{i,j} g^{ij}(-e_j e_i - 2g_{ij}) = -\sum_{i,j} g^{ij} g_{ij} \\ &= -\sum_i \delta_i^i = -(1 + \ldots + 1) = -m. \end{aligned}$$

By means of (i) we also have

$$\sum_{j=1}^m e^j e_j = \sum_{j=1}^m (-e_j e^j - 2) = m - 2m = -m.$$

(v) Follows directly from (iv).
(vi) Follows directly from (iv).
(vii) Similar to (iv).
(viii) Follows directly from (vii).
(ix) Follows directly from (vii). □

Finally we consider the algebra generated by either the covariant basis $(e_j : j = 1, \ldots, m)$ or the contravariant basis $(e^j : j = 1, \ldots, m)$ and we observe that the elements of both bases satisfy the multiplication rules of the complex Clifford algebra $\mathbb{C}_m$:

$$e_j e_k + e_k e_j = -2g_{jk}, \quad 1 \leq j, k \leq m$$

and

$$e^j e^k + e^k e^j = -2g^{jk}, \quad 1 \leq j, k \leq m.$$

A covariant basis for $\mathbb{C}_m$ consists of the elements $e_A = e_{i_1} e_{i_2} \ldots e_{i_h}$ where $A = (i_1, i_2, \ldots, i_h) \subset \{1, \ldots, m\} = M$ is such that $1 \leq i_1 < i_2 < \ldots < i_h \leq m$. Similarly, a contravariant basis for $\mathbb{C}_m$ consists of the elements $e^A = e^{i_1} e^{i_2} \ldots e^{i_h}$ where again $A = (i_1, i_2, \ldots, i_h) \subset M$ is such that $1 \leq i_1 < i_2 < \ldots < i_h \leq m$. In both cases, taking $A = \emptyset$, yields the identity element, i.e., $e_\emptyset = e^\emptyset = 1$.

Hence, any element $\lambda \in \mathbb{C}_m$ may be written as

$$\lambda = \sum_A \lambda_A e_A \quad \text{or as} \quad \lambda = \sum_A \lambda_A e^A \quad \text{with} \quad \lambda_A \in \mathbb{C}.$$

In particular, the space spanned by the covariant basis $(e_j : j = 1, \ldots, m)$ or the contravariant basis $(e^j : j = 1, \ldots, m)$ is called the *subspace of Clifford-vectors.*

Remark 2.18. Note that the real Clifford-vector $\underline{\alpha} = \alpha^j e_j$ may be considered as the inner tensorial product of a contravariant vector α^j with real elements with a covariant vector e_j with Clifford numbers as elements, which yields a tensor of rank 0. So the Clifford-vector $\underline{\alpha}$ is a tensor of rank 0; in fact it is a Clifford number.

In the Clifford algebra $\mathbb{C}_m$, we will frequently use the anti-involution called *Hermitian conjugation*, defined by

$$e_j^\dagger = -e_j, \quad j = 1, \ldots, m$$

and

$$\lambda^\dagger = \Big(\sum_A \lambda_A e_A\Big)^\dagger = \sum_A \lambda_A^c e_A^\dagger,$$

where λ_A^c denotes the complex conjugate of λ_A. Note that in particular for a real Clifford-vector $\underline{\alpha} = \alpha^j e_j$: $\underline{\alpha}^\dagger = -\underline{\alpha}$.

The *Hermitian inner product* on $\mathbb{C}_m$ is defined by

$$(\lambda, \mu) = [\lambda^\dagger \mu]_0 \in \mathbb{C},$$

where $[\lambda]_0$ denotes the scalar part of the Clifford number λ. It follows that for $\lambda \in \mathbb{C}_m$ its *Clifford norm* $|\lambda|_0$ is given by

$$|\lambda|_0^2 = (\lambda, \lambda) = [\lambda^\dagger \lambda]_0.$$

Finally let us examine the Clifford product of the two Clifford-vectors $\underline{\alpha} = \alpha^j e_j$ and $\underline{\beta} = \beta^j e_j$:

$$\begin{aligned}\underline{\alpha}\underline{\beta} &= \alpha^j e_j \beta^k e_k = \alpha^j \beta^k e_j e_k \\ &= \frac{1}{2}\alpha^j \beta^k e_j e_k + \frac{1}{2}\alpha^j \beta^k (-e_k e_j - 2g_{jk}) \\ &= -g_{jk}\alpha^j \beta^k + \frac{1}{2}\alpha^j \beta^k (e_j e_k - e_k e_j).\end{aligned}$$

It is found that this product splits up into a scalar part and a so-called bivector part:

$$\underline{\alpha}\,\underline{\beta} = \underline{\alpha} \bullet \underline{\beta} + \underline{\alpha} \wedge \underline{\beta}$$

with

$$\underline{\alpha} \bullet \underline{\beta} = -g_{jk}\alpha^j \beta^k = \frac{1}{2}(\underline{\alpha}\,\underline{\beta} + \underline{\beta}\,\underline{\alpha}) = \frac{1}{2}\{\underline{\alpha}, \underline{\beta}\}$$

the so-called *inner product* and

$$\underline{\alpha} \wedge \underline{\beta} = \frac{1}{2}\alpha^j \beta^k (e_j e_k - e_k e_j) = \frac{1}{2}\alpha^j \beta^k [e_j, e_k] = \frac{1}{2}[\underline{\alpha}, \underline{\beta}]$$

the so-called *outer product.*

In particular we have that

$$\begin{aligned} e_j \bullet e_k &= -g_{jk}, \quad 1 \leq j,k \leq m \\ e_j \wedge e_j &= 0, \quad j = 1, \ldots, m \end{aligned}$$

and similarly for the contravariant basis elements:

$$\begin{aligned} e^j \bullet e^k &= -g^{jk}, \quad 1 \leq j,k \leq m \\ e^j \wedge e^j &= 0, \quad j = 1, \ldots, m. \end{aligned}$$

The outer product of k different basis vectors is defined recursively

$$e_{i_1} \wedge e_{i_2} \wedge \cdots \wedge e_{i_k} = \frac{1}{2}\big(e_{i_1}(e_{i_2} \wedge \cdots \wedge e_{i_k}) + (-1)^{k-1}(e_{i_2} \wedge \cdots \wedge e_{i_k})e_{i_1}\big).$$

For $k = 0, 1, \ldots, m$ fixed, we then call

$$\mathbb{C}_m^k = \Big\{\lambda \in \mathbb{C}_m \ : \ \lambda = \sum_{|A|=k} \lambda_A \, e_{i_1} \wedge e_{i_2} \wedge \cdots \wedge e_{i_k} \ , \ A = (i_1, i_2, \ldots, i_k)\Big\}$$

the *subspace of k-vectors*; i.e., the space spanned by the outer products of k different basis vectors. Note that the 0-vectors and 1-vectors are simply the scalars and Clifford-vectors; the 2-vectors are also called *bivectors*.

2.3. Embeddings of $\mathbb{R}^m$

By identifying the point $(x^1, \ldots, x^m) \in \mathbb{R}^m$ with the 1-vector $\underline{x}$ given by

$$\underline{x} = x^j e_j,$$

the space $\mathbb{R}^m$ is embedded in the Clifford algebra $\mathbb{C}_m$ as the subspace of 1-vectors $\mathbb{R}^1_m$ of the real Clifford algebra $\mathbb{R}_m$. In the same order of ideas, a point $(x^0, x^1, \ldots, x^m) \in \mathbb{R}^{m+1}$ is identified with a paravector $x = x^0 + \underline{x}$ in $\mathbb{R}^0_m \oplus \mathbb{R}^1_m$.

We will equip $\mathbb{R}^m$ with a metric by defining the scalar product of two basis vectors through their dot product:

$$\langle e_j, e_k \rangle = -e_j \bullet e_k = g_{jk}, \quad 1 \leq j, k \leq m$$

and by putting for two vectors $\underline{x}$ and $\underline{y}$:

$$\langle \underline{x}, \underline{y} \rangle = \langle e_j x^j, e_k y^k \rangle = \langle e_j, e_k \rangle x^j y^k = g_{jk} x^j y^k.$$

Note that in this way $\langle \ . \ , \ . \ \rangle$ is indeed a scalar product, since the tensor g_{jk} is symmetric and positive definite. Also note that in particular

$$\|\underline{x}\|^2 = \langle \underline{x}, \underline{x} \rangle = g_{jk} x^j x^k = -\underline{x} \bullet \underline{x} = -\underline{x}^2 = |\underline{x}|_0^2 .$$

We also introduce *spherical coordinates* in $\mathbb{R}^m$ by:

$$\underline{x} = r\underline{\omega}$$

with

$$r = \|\underline{x}\| = \left(g_{jk} x^j x^k\right)^{1/2} \in [0, +\infty[\quad \text{and} \quad \underline{\omega} \in S^{m-1},$$

where S^{m-1} denotes the *unit sphere* in $\mathbb{R}^m$:

$$S^{m-1} = \left\{\underline{\omega} \in \mathbb{R}^1_m \ ; \ \|\underline{\omega}\|^2 = -\underline{\omega}^2 = g_{jk}\omega^j \omega^k = 1\right\}.$$

Now we introduce a new basis for $\mathbb{R}^m \cong \mathbb{R}^1_m$ consisting of eigenvectors of the matrix $G = (g_{jk})$, associated to the metric tensor g_{jk}. As (g_{jk}) is real-symmetric, there exists an orthogonal matrix $A \in \mathrm{O}(m)$ such that

$$A^T G A = \mathrm{diag}(\lambda_1, \ldots, \lambda_m)$$

with $\lambda_1, \ldots, \lambda_m$ the positive eigenvalues of G. We put

$$(E_j) = (e_j)A.$$

We expect the basis $(E_j \ : \ j = 1, \ldots, m)$ to be orthogonal, since E_j $(j = 1, \ldots, m)$ are eigenvectors of the matrix G.

If (x^j) and (X^j) are the column matrices representing the coordinates of $\underline{x}$ with respect to the bases (e_j) and (E_j) respectively, then we have

$$\underline{x} = (e_j)(x^j) = (E_j)(X^j) = (e_j)A(X^j)$$

and

$$(x^j) = A(X^j) \quad \text{or} \quad (X^j) = A^T(x^j).$$

Hence

$$\begin{aligned}\langle \underline{x}, \underline{y} \rangle &= (x^j)^T G(y^j) = (X^j)^T A^T G A (Y^j) \\ &= (X^j)^T \operatorname{diag}(\lambda_1, \ldots, \lambda_m)\ (Y^j) = \sum_j \lambda_j X^j Y^j.\end{aligned}$$

In particular we find, as expected:

$$\langle E_j, E_k \rangle = 0, \quad j \neq k \quad \text{and} \quad \langle E_j, E_j \rangle = \lambda_j, \quad j = 1, \ldots, m.$$

Involving the Clifford product we obtain

$$\{E_j, E_k\} = 2E_j \bullet E_k = -2\langle E_j, E_k \rangle = 0, \quad j \neq k$$

and

$$\{E_j, E_j\} = 2E_j \bullet E_j = -2\langle E_j, E_j \rangle = -2\lambda_j, \quad j = 1, \ldots, m$$

and so for all $1 \leq j, k \leq m$

$$E_j E_k + E_k E_j = -2\lambda_j \delta_{jk}.$$

Finally if we put

$$\underline{\eta}_j = \frac{E_j}{\sqrt{\lambda_j}} \quad (j = 1, \ldots, m),$$

we obtain an orthonormal frame in the metric dependent setting:

$$\langle \underline{\eta}_j, \underline{\eta}_k \rangle = \delta_{jk}; \quad 1 \leq j, k \leq m.$$

In what follows, (x'^j) denotes the column matrix containing the coordinates of $\underline{x}$ with respect to this orthonormal frame, i.e.,

$$\underline{x} = \sum_j x'^j\ \underline{\eta}_j = \sum_j \langle \underline{\eta}_j, \underline{x} \rangle\ \underline{\eta}_j.$$

It is clear that the coordinate sets (X^j) and (x'^j) are related as follows:

$$(X^j) = P^{-1}(x'^j) \quad \text{or} \quad (x'^j) = P(X^j)$$

with

$$P = \operatorname{diag}(\sqrt{\lambda_1}, \ldots, \sqrt{\lambda_m}).$$

Hence we also have

$$(x^j) = AP^{-1}(x'^j) \quad \text{and} \quad (x'^j) = PA^T(x^j).$$

Finally, note that in the x'^j-coordinates, the inner product takes the following form

$$\langle \underline{x}, \underline{y} \rangle = \sum_j x'^j y'^j.$$

2.4. Fischer duality and Fischer decomposition

We consider the algebra $\mathcal{P}$ of Clifford algebra-valued polynomials, generated by $\{x^1, \ldots, x^m \,;\, e_1, \ldots, e_m; i\}$. A natural inner product on $\mathcal{P}$ is the so-called *Fischer inner product*

$$(R(\underline{x}), S(\underline{x})) = \left[\left\{ R^\dagger (g^{1j}\partial_{x^j}, g^{2j}\partial_{x^j}, \ldots, g^{mj}\partial_{x^j})[S(\underline{x})] \right\}_{\underline{x}=\underline{0}} \right]_0 .$$

Note that in order to obtain the differential operator $R^\dagger(g^{1j}\partial_{x^j}, \ldots, g^{mj}\partial_{x^j})$, one first takes the Hermitian conjugate $R^\dagger$ of the polynomial R, followed by the substitution $x^k \to g^{kj}\partial_{x^j}$. These two operations

$$\begin{cases} e_k & \xrightarrow{F} \quad -e_k \quad \text{(Hermitian conjugation)} \\ x^k & \xrightarrow{F} \quad g^{kj}\partial_{x^j} \end{cases}$$

are known as *Fischer duality*.

Now we express this Fischer duality in terms of the new basis $(E_j \ : \ j = 1, \ldots, m)$ and the corresponding new coordinates (X^j) introduced in the foregoing subsection:

(i) $(E_j) = (e_j)A \xrightarrow{F} (-e_j)A = -(e_j)A = -(E_j)$ (Hermitian conjugation)

(ii)
$$\begin{aligned} X^j = A^T_{jk}x^k \xrightarrow{F} A^T_{jk}g^{ik}\partial_{x^i} &= A^T_{jk}g^{ik}\frac{\partial X^\ell}{\partial x^i}\partial_{X^\ell} = A^T_{jk}g^{ik}A^T_{\ell i}\partial_{X^\ell} \\ &= A^T_{jk}g^{ik}A_{i\ell}\,\partial_{X^\ell} = (A^T G^{-1} A)_{j\ell}\,\partial_{X^\ell} \\ &= \left(\operatorname{diag}\left(\frac{1}{\lambda_1}, \ldots, \frac{1}{\lambda_m}\right)\right)_{j\ell} \partial_{X^\ell} = \frac{1}{\lambda_j}\partial_{X^j}. \quad (2.3) \end{aligned}$$

Proposition 2.19. *The basis* $\left(E_A \underline{X}^{\underline{\alpha}} = E_A (X^1)^{\alpha_1} \ldots (X^m)^{\alpha_m} : A \subset M, \underline{\alpha} \in \mathbb{N}^m\right)$ *of the space* $\mathcal{P}$ *of Clifford polynomials is orthogonal with respect to the Fischer inner product.*

Proof. We have consecutively

$$\begin{aligned} &(E_A\underline{X}^{\underline{\alpha}}, E_B\underline{X}^{\underline{\beta}}) \\ &= \left[E_A^\dagger \left(\frac{1}{\lambda_1}\partial_{X^1}\right)^{\alpha_1} \cdots \left(\frac{1}{\lambda_m}\partial_{X^m}\right)^{\alpha_m} [E_B\underline{X}^{\underline{\beta}}] \right]_0 /_{\underline{X}=\underline{0}} \\ &= \left[E_A^\dagger E_B \left(\frac{1}{\lambda_1}\right)^{\alpha_1} \cdots \left(\frac{1}{\lambda_m}\right)^{\alpha_m} (\partial_{X^1})^{\alpha_1} \ldots (\partial_{X^m})^{\alpha_m} (X^1)^{\beta_1} \ldots (X^m)^{\beta_m} \right]_0 /_{\underline{X}=\underline{0}} \\ &= \left(\frac{1}{\lambda_1}\right)^{\alpha_1} \cdots \left(\frac{1}{\lambda_m}\right)^{\alpha_m} \underline{\alpha}!\ \delta_{\underline{\alpha},\underline{\beta}}\ [E_A^\dagger E_B]_0 \end{aligned}$$

with $\underline{\alpha}! = \alpha_1! \ \ldots \ \alpha_m!$. Moreover, for

$$E_A = E_{i_1}E_{i_2}\ldots E_{i_h} \quad \text{and} \quad E_B = E_{j_1}E_{j_2}\ldots E_{j_k},$$

we find

$$E_A^\dagger E_B = (-1)^{|A|}\ E_{i_h}\ldots E_{i_2}E_{i_1}E_{j_1}E_{j_2}\ldots E_{j_k}.$$

As

$$E_j E_k + E_k E_j = -2\lambda_j \delta_{jk}, \quad 1 \leq j,k \leq m,$$

we have

$$\begin{aligned}[E_A^\dagger E_B]_0 &= (-1)^{|A|}\ (-\lambda_{i_1})(-\lambda_{i_2})\ldots(-\lambda_{i_h})\ \delta_{A,B} \\ &= \lambda_{i_1}\lambda_{i_2}\ldots\lambda_{i_h}\ \delta_{A,B}.\end{aligned}$$

Summarizing, we have found that for $A = (i_1, i_2, \ldots, i_h)$

$$(E_A \underline{X}^{\underline{\alpha}}, E_B \underline{X}^{\underline{\beta}}) = \underline{\alpha}!\ \lambda_{i_1}\lambda_{i_2}\ldots\lambda_{i_h} \left(\frac{1}{\lambda_1}\right)^{\alpha_1} \cdots \left(\frac{1}{\lambda_m}\right)^{\alpha_m} \delta_{A,B}\ \delta_{\underline{\alpha},\underline{\beta}}. \qquad \square$$

Proposition 2.20. *The Fischer inner product is positive definite.*

Proof. This follows in a straightforward way from the fact that the Fischer inner product of a basis polynomial $E_A \underline{X}^{\underline{\alpha}}$ with itself is always positive:

$$\left(E_A \underline{X}^{\underline{\alpha}}, E_A \underline{X}^{\underline{\alpha}}\right) = \underline{\alpha}!\ \lambda_{i_1}\lambda_{i_2}\ldots\lambda_{i_h} \left(\frac{1}{\lambda_1}\right)^{\alpha_1} \cdots \left(\frac{1}{\lambda_m}\right)^{\alpha_m} > 0. \qquad \square$$

Let $\mathcal{P}_k$ denote the subspace of $\mathcal{P}$ consisting of the homogeneous Clifford polynomials of degree k:

$$\mathcal{P}_k = \{R_k(\underline{x}) \in \mathcal{P}\ :\ R_k(t\underline{x}) = t^k R_k(\underline{x})\ ,\ t \in \mathbb{R}\}.$$

It follows from Proposition 2.19 that the spaces $\mathcal{P}_k$ are orthogonal with respect to the Fischer inner product. With a view to the *Fischer decomposition* of the homogeneous Clifford polynomials, we now introduce the notion of *monogenicity*, which in fact is at the heart of Clifford analysis in the same way as the notion of holomorphicity is fundamental to the function theory in the complex plane. Monogenicity is defined by means of the so-called *Dirac-operator* $\partial_{\underline{x}}$ which we introduce as the Fischer dual of the vector variable $\underline{x}$:

$$\underline{x} = x^j e_j \quad \xrightarrow{F} \quad -g^{jk}\partial_{x^k} e_j = -e^k \partial_{x^k} = -\partial_{\underline{x}}\ .$$

This Dirac operator factorizes the *g-Laplacian*, which we obtain as the Fischer dual of the scalar function $\underline{x}^2 = -\|\underline{x}\|^2$:

$$\begin{aligned}\underline{x}^2 = -\langle \underline{x}, \underline{x}\rangle = -g_{jk}x^j x^k \xrightarrow{F} -g_{jk}g^{ji}\partial_{x^i}g^{k\ell}\partial_{x^\ell} = -g_{jk}g^{ji}g^{k\ell}\partial_{x^i}\partial_{x^\ell} = -\delta^i_k g^{k\ell}\partial_{x^i}\partial_{x^\ell} \\ = -g^{i\ell}\partial^2_{x^i x^\ell} = -\Delta_g,\end{aligned}$$

where we have defined the g-Laplacian as to be

$$\Delta_g = g^{jk}\partial^2_{x^j x^k}.$$

Then we have indeed that

$$\begin{aligned}\partial^2_{\underline{x}} &= \partial_{x^j} e^j \partial_{x^k} e^k = \frac{1}{2}\partial_{x^j}\partial_{x^k} e^j e^k + \frac{1}{2}\partial_{x^k}\partial_{x^j} e^k e^j \\ &= \frac{1}{2}\partial_{x^j}\partial_{x^k} e^j e^k + \frac{1}{2}\partial_{x^j}\partial_{x^k}(-e^j e^k - 2g^{jk}) \\ &= \frac{1}{2}\partial_{x^j}\partial_{x^k} e^j e^k - \frac{1}{2}\partial_{x^j}\partial_{x^k} e^j e^k - g^{jk}\partial_{x^j}\partial_{x^k} = -\Delta_g.\end{aligned}$$

We also mention the expression for the Dirac operator in the orthonormal frame introduced at the end of Subsection 2.3.

Lemma 2.21. *With respect to the orthonormal frame $\underline{\eta}_j$ $(j = 1, \ldots, m)$, the Dirac operator takes the form*

$$\partial_{\underline{x}} = \sum_j \partial_{x'^j} \underline{\eta}_j.$$

Proof. For the sake of clarity, we do not use the Einstein summation convention. We have consecutively

$$\begin{aligned}\partial_{\underline{x}} &= \sum_j \partial_{x^j} e^j = \sum_{j,k} \partial_{x^j} g^{jk} e_k = \sum_{j,k,t} \partial_{x^j} g^{jk} E_t A^T_{tk} \\ &= \sum_t \left(\sum_{j,k} A^T_{tk} g^{jk} \partial_{x^j} \right) E_t.\end{aligned}$$

By means of (2.3) this becomes

$$\partial_{\underline{x}} = \sum_t \frac{1}{\lambda_t} \partial_{X^t} E_t = \sum_t \partial_{x'^t} \underline{\eta}_t. \qquad \square$$

Definition 2.22.

(i) A $\mathbb{C}_m$-valued function $F(x^1, \ldots, x^m)$ is called *left monogenic* in an open region of $\mathbb{R}^m$ if in that region

$$\partial_{\underline{x}}[F] = 0.$$

(ii) A $\mathbb{C}_m$-valued function $F(x^0, x^1, \ldots, x^m)$ is called *left monogenic* in an open region of $\mathbb{R}^{m+1}$ if in that region

$$(\partial_{x^0} + \partial_{\underline{x}})[F] = 0.$$

Here $\partial_{x^0} + \partial_{\underline{x}}$ is the so-called *Cauchy-Riemann operator.*

The notion of *right monogenicity* is defined in a similar way by letting act the Dirac operator or the Cauchy-Riemann operator from the right.

Note that if a Clifford algebra-valued function F is left monogenic, then its Hermitian conjugate $F^\dagger$ is right monogenic, since $\partial^\dagger_{\underline{x}} = -\partial_{\underline{x}}$.

Similar to the notion of spherical harmonic with respect to the Laplace operator we introduce the following fundamental concept.

Definition 2.23.

(i) A (left/right) monogenic homogeneous polynomial $P_k \in \mathcal{P}_k$ is called a (left/right) *solid inner spherical monogenic* of order k.

(ii) A (left/right) monogenic homogeneous function Q_k of degree $-(k + m - 1)$ in $\mathbb{R}^m \setminus \{0\}$ is called a (left/right) *solid outer spherical monogenic* of order k.

The set of all left, respectively right, solid inner spherical monogenics of order k will be denoted by $M_\ell^+(k)$, respectively $M_r^+(k)$, while the set of all left, respectively right, solid outer spherical monogenics of order k will be denoted by $M_\ell^-(k)$, respectively $M_r^-(k)$.

Theorem 2.24 (Fischer decomposition).

(i) *Any $R_k \in \mathcal{P}_k$ has a unique orthogonal decomposition of the form*

$$R_k(\underline{x}) = P_k(\underline{x}) + \underline{x} R_{k-1}(\underline{x})$$

with $P_k \in M_\ell^+(k)$ and $R_{k-1} \in \mathcal{P}_{k-1}$.

(ii) *The space of homogeneous polynomials $\mathcal{P}_k$ admits the orthogonal decomposition: $\mathcal{P}_k = M_\ell^+(k) \oplus_\perp \underline{x}\mathcal{P}_{k-1}$.*

Proof. This orthogonal decomposition follows from the observation that for $R_{k-1} \in \mathcal{P}_{k-1}$ and $S_k \in \mathcal{P}_k$:

$$(\underline{x} R_{k-1}(\underline{x}), S_k(\underline{x})) = -(R_{k-1}(\underline{x}), \partial_{\underline{x}}[S_k(\underline{x})]). \tag{2.4}$$

Indeed, as $-\partial_{\underline{x}}$ is the Fischer dual of $\underline{x}$, we have

$$\begin{aligned}(\underline{x} R_{k-1}(\underline{x}), S_k(\underline{x})) &= -\left[R_{k-1}^\dagger \left(g^{1j}\partial_{x^j}, \ldots, g^{mj}\partial_{x^j} \right) \partial_{\underline{x}}[S_k(\underline{x})] \right]_0 \\ &= -\left(R_{k-1}(\underline{x}), \partial_{\underline{x}}[S_k(\underline{x})] \right).\end{aligned}$$

Next, if for some $S_k \in \mathcal{P}_k$ and for all $R_{k-1} \in \mathcal{P}_{k-1}$

$$(\underline{x} R_{k-1}, S_k) = 0,$$

then so will

$$(R_{k-1}, \partial_{\underline{x}}[S_k]) = 0.$$

Hence $\partial_{\underline{x}}[S_k] = 0$, which means that the orthogonal complement of $\underline{x}\mathcal{P}_{k-1}$ is a subspace of $M_\ell^+(k)$.

But if $P_k = \underline{x} R_{k-1} \in M_\ell^+(k) \cap \underline{x}\mathcal{P}_{k-1}$, then

$$(P_k, P_k) = (\underline{x} R_{k-1}, P_k) = -(R_{k-1}, \partial_{\underline{x}}[P_k]) = 0$$

and thus $P_k = 0$. So any $S_k \in \mathcal{P}_k$ may be uniquely decomposed as

$$S_k = P_k + \underline{x} R_{k-1} \quad \text{with} \quad P_k \in M_\ell^+(k) \quad \text{and} \quad R_{k-1} \in \mathcal{P}_{k-1}.$$

□

Theorem 2.25 (monogenic decomposition). *Any $R_k \in \mathcal{P}_k$ has a unique orthogonal decomposition of the form*

$$R_k(\underline{x}) = \sum_{s=0}^{k} \underline{x}^s P_{k-s}(\underline{x}), \quad \text{with} \quad P_{k-s} \in M_\ell^+(k-s).$$

Proof. This result follows by recursive application of the Fischer decomposition.

□

2.5. The Euler and angular Dirac operators

The Euler and angular Dirac operators are two fundamental operators arising quite naturally when considering the operator $\underline{x}\partial_{\underline{x}}$; in fact they are the respective scalar and bivector part of it.

Definition 2.26. The *Euler operator* is the operator defined by

$$E = x^j \partial_{x^j};$$

the *angular Dirac operator* is defined by

$$\Gamma = -\frac{1}{2} x^j \partial_{x^k} (e_j e^k - e^k e_j) = -x^j \partial_{x^k} e_j \wedge e^k.$$

Proposition 2.27. *One has*

$$\underline{x}\partial_{\underline{x}} = -E - \Gamma \tag{2.5}$$

or in other words

$$-E = \underline{x} \bullet \partial_{\underline{x}} = [\underline{x}\partial_{\underline{x}}]_0 \quad \text{and} \quad -\Gamma = \underline{x} \wedge \partial_{\underline{x}} = [\underline{x}\partial_{\underline{x}}]_2,$$

where $[\lambda]_2$ *denotes the bivector part of the Clifford number* λ.

Proof. One easily finds

$$\begin{aligned}
\underline{x}\partial_{\underline{x}} = x^j e_j \partial_{x^k} e^k &= \frac{1}{2} x^j \partial_{x^k} e_j e^k + \frac{1}{2} x^j \partial_{x^k} e_j e^k \\
&= \frac{1}{2} x^j \partial_{x^k} e_j e^k + \frac{1}{2} x^j \partial_{x^k} (-e^k e_j - 2\delta_j^k) \\
&= -x^j \partial_{x^j} + \frac{1}{2} x^j \partial_{x^k} (e_j e^k - e^k e_j) = -E - \Gamma.
\end{aligned}$$

□

As is well known, the Euler operator measures the degree of homogeneity of polynomials, while the angular Dirac operator measures the degree of monogenicity. This is expressed by the following eigenvalue equations.

Proposition 2.28.

(i) *For* $R_k \in \mathcal{P}_k$ *one has*

$$E[R_k] = kR_k.$$

(ii) *For* $P_k \in M_\ell^+(k)$ *one has*

$$\Gamma[P_k] = -kP_k.$$

Proof. (i) A homogeneous polynomial R_k of degree k can be written as

$$R_k(\underline{x}) = \sum_A e_A \, R_{k,A}(\underline{x})$$

with $R_{k,A}$ a scalar-valued homogeneous polynomial of degree k. Hence

$$E[R_k] = \sum_A e_A \, E[R_{k,A}] = \sum_A e_A k R_{k,A} = kR_k.$$

(ii) Using (2.5) it is easily seen that

$$\Gamma[P_k] = (-E - \underline{x}\partial_{\underline{x}})[P_k] = -kP_k.$$

□

Remark 2.29. A Clifford polynomial operator is an element of $\mathrm{End}(\mathcal{P})$; it transforms a Clifford polynomial into another one. Such a Clifford polynomial operator A_ℓ is called homogeneous of degree ℓ if

$$A_\ell[\mathcal{P}_k] \subset \mathcal{P}_{k+\ell}.$$

The Euler operator also measures the degree of homogeneity of Clifford polynomial operators.

Proposition 2.30. *The Clifford polynomial operator A_ℓ is homogeneous of degree ℓ if and only if*

$$[E, A_\ell] = \ell A_\ell.$$

Proof. Suppose that A_ℓ is homogeneous of degree ℓ. We then have

$$(EA_\ell - A_\ell E)[\mathcal{P}_k] = E[\mathcal{P}_{k+\ell}] - kA_\ell[\mathcal{P}_k] = (k+\ell)\mathcal{P}_{k+\ell} - k\mathcal{P}_{k+\ell} = \ell A_\ell[\mathcal{P}_k].$$

Conversely, assume that the Clifford polynomial operator B satisfies

$$[E, B] = \lambda B.$$

For an arbitrary $R_k \in \mathcal{P}_k$ we then have

$$E(B[R_k]) = (BE + \lambda B)[R_k] = kB[R_k] + \lambda B[R_k] = (k+\lambda)B[R_k].$$

In other words, $B[R_k] \in \mathcal{P}_{k+\lambda}$ and hence B is homogeneous of degree λ. □

In a series of lemmata and propositions we now establish the basic formulae and operator identities needed in the sequel.

Lemma 2.31. *One has*

$$\partial_{\underline{x}}[\underline{x}] = -m.$$

Proof. By means of Proposition 2.17, we have immediately

$$\partial_{\underline{x}}[\underline{x}] = \sum_{k,j} e^k \partial_{x^k}(x^j) e_j = \sum_j e^j e_j = -m.$$ □

Proposition 2.32. *One has*

$$\underline{x}\partial_{\underline{x}} + \partial_{\underline{x}}\underline{x} = -2E - m \quad \text{and} \quad \partial_{\underline{x}}\underline{x} = -E - m + \Gamma. \tag{2.6}$$

Proof. On the one hand we have

$$\underline{x}\partial_{\underline{x}} = \sum_{j,k} x^j e_j \partial_{x^k} e^k,$$

while on the other hand

$$\partial_{\underline{x}}\underline{x} = \partial_{\underline{x}}[\underline{x}] + \dot{\partial}_{\underline{x}}\underline{x} = -m + \sum_{j,k} \dot{\partial}_{x^k} e^k x^j e_j = -m + \sum_{j,k} x^j \partial_{x^k} e^k e_j,$$

where the dot-notation $\dot{\partial}_{\underline{x}}$ means that the Dirac operator does not act on the function $\underline{x}$, but on the function at the right of it.

Adding both equalities yields

$$\begin{aligned}\underline{x}\partial_{\underline{x}} + \partial_{\underline{x}}\underline{x} &= -m + \sum_{j,k} x^j \partial_{x^k}(e_j e^k + e^k e_j)\\ &= -m - 2\sum_{j,k} x^j \partial_{x^k} \delta_j^k = -m - 2E.\end{aligned}$$

Using (2.5) this becomes

$$\partial_{\underline{x}}\underline{x} = -\underline{x}\partial_{\underline{x}} - 2E - m = E + \Gamma - 2E - m = \Gamma - E - m. \qquad \square$$

Propositions 2.27 and 2.32 yield some nice additional results. The first corollary focuses on an interesting factorization of the g-Laplacian.

Corollary 2.33. *One has*

$$\Delta_g = (E + m - \Gamma)\ \frac{1}{||\underline{x}||^2}\ (E + \Gamma).$$

Proof. By means of (2.5) and (2.6), we obtain

$$\begin{aligned}\Delta_g &= -\partial_{\underline{x}}^2 = \partial_{\underline{x}} \frac{\underline{x}^2}{||\underline{x}||^2} \partial_{\underline{x}} = \partial_{\underline{x}}\ \underline{x}\ \frac{1}{||\underline{x}||^2}\ \underline{x}\ \partial_{\underline{x}}\\ &= (E + m - \Gamma)\frac{1}{||\underline{x}||^2}(E + \Gamma).\end{aligned} \qquad \square$$

Also the polynomials $\underline{x}P_k$ are eigenfunctions of the angular Dirac operator.

Corollary 2.34. *For any $P_k \in M_\ell^+(k)$ one has:*

$$\Gamma[\underline{x}P_k] = (k + m - 1)\underline{x}P_k.$$

Proof. By means of (2.6) we obtain

$$\begin{aligned}\underline{x}\partial_{\underline{x}}[\underline{x}P_k] &= \underline{x}(-E - m + \Gamma)[P_k]\\ &= \underline{x}(-k - m - k)P_k = (-2k - m)\underline{x}P_k.\end{aligned}$$

Using (2.5) and the fact that $\underline{x}P_k \in \mathcal{P}_{k+1}$ gives

$$(-2k - m)\underline{x}P_k = -(E + \Gamma)[\underline{x}P_k] = -(k + 1)\underline{x}P_k - \Gamma[\underline{x}P_k].$$

Hence

$$\Gamma[\underline{x}P_k] = (k + m - 1)\underline{x}P_k. \qquad \square$$

Lemma 2.35. *One has*

(i)
$$\partial_{x^j}[r] = \frac{1}{r} g_{jk} x^k \quad , \quad j = 1, \ldots, m$$

(ii)
$$\partial_{\underline{x}}[r] = \frac{\underline{x}}{r}.$$

Proof. A straightforward computation yields

(i) $$\partial_{x^j}[r] = \partial_{x^j}\left[\left(g_{ik}x^i x^k\right)^{1/2}\right] = \frac{1}{2}\left(g_{ik}x^i x^k\right)^{-1/2}\left(g_{jk}x^k + g_{ij}x^i\right) = \frac{1}{2}\frac{1}{r}2g_{jk}x^k = \frac{1}{r}g_{jk}x^k$$

(ii) $$\partial_{\underline{x}}[r] = e^j\partial_{x^j}[r] = \frac{1}{r}e^j g_{jk}x^k = \frac{1}{r}e_k x^k = \frac{\underline{x}}{r}.$$ □

Now it is easily proved that the angular Dirac operator only acts on the angular coordinates, whence its name.

Lemma 2.36. *One has*

(i) $E[r] = r$
(ii) $\underline{x}\partial_{\underline{x}}[r] = -r$
(iii) $\Gamma[r] = 0$.

Proof. By means of the previous Lemma we easily find

(i) $$E[r] = x^j\partial_{x^j}[r] = x^j\frac{1}{r}g_{jk}x^k = \frac{1}{r}r^2 = r.$$

(ii) $$\underline{x}\partial_{\underline{x}}[r] = \underline{x}\frac{\underline{x}}{r} = -\frac{r^2}{r} = -r.$$

(iii) $$\Gamma[r] = -(\underline{x}\partial_{\underline{x}} + E)[r] = -(-r+r) = 0.$$ □

This enables us to prove the following eigenvalue equations.

Theorem 2.37. *One has*

(i) $E[\underline{x}^s P_k] = (s+k)\underline{x}^s P_k$
(ii) $\Gamma[\underline{x}^{2s}P_k] = -k\underline{x}^{2s}P_k$
(iii) $\Gamma[\underline{x}^{2s+1}P_k] = (k+m-1)\underline{x}^{2s+1}P_k$.

Proof. (i) Follows immediately from the fact that $\underline{x}^s P_k \in \mathcal{P}_{s+k}$.
(ii) By means of the previous Lemma we find

$$\Gamma[\underline{x}^{2s}P_k] = \underline{x}^{2s}\Gamma[P_k] = -k\underline{x}^{2s}P_k.$$

(iii) Similarly we have

$$\Gamma[\underline{x}^{2s+1}P_k] = \underline{x}^{2s}\Gamma[\underline{x}P_k] = (k+m-1)\underline{x}^{2s+1}P_k.$$ □

The previous Theorem combined with (2.6) yields the following result.

Theorem 2.38. *One has*

$$\partial_{\underline{x}}[\underline{x}^s P_k] = B_{s,k}\ \underline{x}^{s-1}P_k$$

with

$$B_{s,k} = \begin{cases} -s & \text{for } s \text{ even,} \\ -(s-1+2k+m) & \text{for } s \text{ odd.} \end{cases}$$

Proof. We have consecutively

$$\begin{aligned}\partial_{\underline{x}}[\underline{x}^{2s}P_k] &= \partial_{\underline{x}}\underline{x}[\underline{x}^{2s-1}P_k] = (-E-m+\Gamma)[\underline{x}^{2s-1}P_k] \\ &= -(2s+k-1)\underline{x}^{2s-1}P_k - m\underline{x}^{2s-1}P_k + (k+m-1)\underline{x}^{2s-1}P_k \\ &= -2s\underline{x}^{2s-1}P_k\end{aligned}$$

and similarly

$$\begin{aligned}\partial_{\underline{x}}[\underline{x}^{2s+1}P_k] &= \partial_{\underline{x}}\underline{x}[\underline{x}^{2s}P_k] = (-E-m+\Gamma)[\underline{x}^{2s}P_k] \\ &= -(2s+k)\underline{x}^{2s}P_k - m\underline{x}^{2s}P_k - k\underline{x}^{2s}P_k \\ &= -(2s+2k+m)\underline{x}^{2s}P_k.\end{aligned}$$

□

Solid inner and outer spherical monogenics are related to each other by means of the so-called spherical transform.

Proposition 2.39. *If P_k is a solid inner spherical monogenic of degree k, then*

$$Q_k(\underline{x}) = \frac{\underline{x}}{||\underline{x}||^m}\, P_k\left(\frac{\underline{x}}{||\underline{x}||^2}\right) = \frac{\underline{x}}{||\underline{x}||^{2k+m}}\, P_k(\underline{x})$$

is a monogenic homogeneous function of degree $-(k+m-1)$ in $\mathbb{R}^m \setminus \{0\}$, i.e., a solid outer spherical monogenic of degree k.

Conversely, if $\widetilde{Q}_k$ is a solid outer spherical monogenic of degree k, then

$$\widetilde{P}_k(\underline{x}) = \frac{\underline{x}}{||\underline{x}||^m}\, \widetilde{Q}_k\left(\frac{\underline{x}}{||\underline{x}||^2}\right) = \underline{x}\, ||\underline{x}||^{2k+m-2}\, \widetilde{Q}_k(\underline{x})$$

is a solid inner spherical monogenic of degree k.

Proof. Clearly Q_k is homogeneous of degree $-(k+m-1)$, since

$$Q_k(t\underline{x}) = \frac{t^{k+1}\underline{x}P_k(\underline{x})}{t^{2k+m}||\underline{x}||^{2k+m}} = t^{-(k+m-1)}Q_k(\underline{x}).$$

Moreover, in $\mathbb{R}^m \setminus \{0\}$:

$$\begin{aligned}\partial_{\underline{x}}[Q_k(\underline{x})] &= \partial_{\underline{x}}\left[\frac{1}{||\underline{x}||^{2k+m}}\right]\underline{x}P_k + \frac{1}{||\underline{x}||^{2k+m}}\partial_{\underline{x}}[\underline{x}P_k] \\ &= -(2k+m)\frac{\underline{x}}{||\underline{x}||^{2k+m+2}}\underline{x}P_k - (2k+m)\frac{1}{||\underline{x}||^{2k+m}}P_k \\ &= \frac{(2k+m)}{||\underline{x}||^{2k+m}}P_k - \frac{(2k+m)}{||\underline{x}||^{2k+m}}P_k = 0,\end{aligned}$$

where we have used

$$\partial_{\underline{x}}\left[\frac{1}{r^{2k+m}}\right] = -(2k+m)\frac{1}{r^{2k+m+1}}\partial_{\underline{x}}[r] = -\frac{(2k+m)}{r^{2k+m+1}}\frac{\underline{x}}{r} = -(2k+m)\frac{\underline{x}}{r^{2k+m+2}}.$$

The converse result is proved in a similar way. □

Proposition 2.40. *In terms of spherical coordinates the Euler operator E takes the form*

$$E = r\partial_r .$$

Proof. In this proof we do not use the Einstein summation convention.

The transformation formulae from cartesian to spherical coordinates in $\mathbb{R}^m$ yield

$$\partial_{x^j} = \frac{\partial r}{\partial x^j}\partial_r + \sum_{k=1}^{m} \frac{\partial \omega^k}{\partial x^j}\partial_{\omega^k}. \tag{2.7}$$

In view of Lemma 2.35 we have for each j fixed:

$$\frac{\partial r}{\partial x^j} = \frac{1}{r}\sum_t g_{jt}x^t,$$

and

$$\begin{aligned}\frac{\partial \omega^j}{\partial x^j} &= \frac{\partial}{\partial x^j}\left(\frac{x^j}{r}\right) = \frac{1}{r} - x^j\frac{1}{r^2}\frac{\partial r}{\partial x^j}\\ &= \frac{1}{r} - \frac{1}{r^3}x^j\sum_t g_{jt}x^t = \frac{1}{r}\big(1 - \omega^j \sum_t g_{jt}\omega^t\big),\end{aligned}$$

while for $k \neq j$, we find

$$\begin{aligned}\frac{\partial \omega^k}{\partial x^j} &= \frac{\partial}{\partial x^j}\left(\frac{x^k}{r}\right) = -x^k\frac{1}{r^2}\frac{\partial r}{\partial x^j}\\ &= -x^k\frac{1}{r^3}\sum_t g_{jt}x^t = -\omega^k\frac{1}{r}\sum_t g_{jt}\omega^t.\end{aligned}$$

Hence equation (2.7) becomes

$$\begin{aligned}\partial_{x^j} &= \frac{1}{r}\big(\sum_t g_{jt}x^t\big)\partial_r + \frac{1}{r}\big(1-\omega^j\big(\sum_t g_{jt}\omega^t\big)\big)\partial_{\omega^j} - \sum_{k=1,\ k\neq j}^{m} \omega^k\frac{1}{r}\big(\sum_t g_{jt}\omega^t\big)\partial_{\omega^k}\\ &= \frac{1}{r}\big(\sum_t g_{jt}x^t\big)\partial_r + \frac{1}{r}\partial_{\omega^j} - \frac{1}{r}\sum_{k=1}^{m}\omega^k\big(\sum_t g_{jt}\omega^t\big)\partial_{\omega^k}. \end{aligned}\tag{2.8}$$

Thus we finally obtain

$$\begin{aligned}E &= \sum_j x^j\partial_{x^j} = \frac{1}{r}\sum_j x^j\big(\sum_t g_{jt}x^t\big)\partial_r + \frac{1}{r}\sum_j x^j\partial_{\omega^j} - \frac{1}{r}\sum_j x^j\sum_k \omega^k\big(\sum_t g_{jt}\omega^t\big)\partial_{\omega^k}\\ &= \frac{1}{r}r^2\partial_r + \sum_j \omega^j\partial_{\omega^j} - \sum_j\omega^j\sum_k\omega^k\big(\sum_t g_{jt}\omega^t\big)\partial_{\omega^k}\\ &= r\partial_r + \sum_j\omega^j\partial_{\omega^j} - \sum_k\omega^k\sum_{j,t}g_{jt}\omega^j\omega^t\partial_{\omega^k}\\ &= r\partial_r + \sum_j\omega^j\partial_{\omega^j} - \sum_k\omega^k\partial_{\omega^k} = r\partial_r.\end{aligned}$$

□

The foregoing Proposition 2.40 enables us to express also the Dirac operator in spherical coordinates.

Proposition 2.41. *In terms of spherical coordinates the Dirac operator $\partial_{\underline{x}}$ takes the form*

$$\partial_{\underline{x}} = \underline{\omega} \left(\partial_r + \frac{1}{r}\Gamma \right).$$

Proof. By means of (2.5) and Proposition 2.40 we find

$$r\underline{\omega}\partial_{\underline{x}} = -(r\partial_r + \Gamma) \qquad \text{or} \qquad \underline{\omega}\partial_{\underline{x}} = -\left(\partial_r + \frac{\Gamma}{r} \right).$$

Multiplying both sides of the above equation by $-\underline{\omega}$ yields the desired result. □

Lemma 2.42. *One has*

$$\Gamma[\underline{\omega}] = (m-1)\underline{\omega}, \quad \underline{\omega} \in S^{m-1}.$$

Proof. Combining Lemma 2.31 with the previous Proposition 2.41 yields

$$\underline{\omega} \left(\partial_r + \frac{\Gamma}{r} \right) [r\underline{\omega}] = -m.$$

Hence we also have

$$\underline{\omega}(\underline{\omega} + \Gamma[\underline{\omega}]) = -m \qquad \text{or} \qquad \underline{\omega}\Gamma[\underline{\omega}] = -(m-1).$$

Multiplying both sides of the above equation by $-\underline{\omega}$ yields the desired result. □

We end this subsection with establishing some commutation relations involving the angular Dirac operator.

Proposition 2.43. *One has*

(i) $\Gamma\underline{\omega} = \underline{\omega}(m-1-\Gamma)$
(ii) $\Gamma\underline{x} = \underline{x}(m-1-\Gamma)$
(iii) $\Gamma\partial_{\underline{x}} = \partial_{\underline{x}}(m-1-\Gamma)$.

Proof. (i) We have

$$(\Gamma\underline{\omega})f = \Gamma[\underline{\omega}f] = \Gamma[\underline{\omega}]f + \dot{\Gamma}\underline{\omega}\dot{f} = (m-1)\underline{\omega}f + \dot{\Gamma}\underline{\omega}\dot{f}.$$

Using (2.8) yields

$$\begin{aligned}
\partial_{\underline{x}} &= \sum_j e^j \partial_{x^j} = \sum_j e^j (\sum_t g_{jt}\omega^t)\partial_r + \frac{1}{r}\sum_j e^j \partial_{\omega^j} - \frac{1}{r}\sum_j e^j \sum_k \omega^k (\sum_t g_{jt}\omega^t)\partial_{\omega^k} \\
&= (\sum_t e_t\omega^t)\partial_r - \frac{\underline{\omega}}{r}\underline{\omega}\sum_j e^j\partial_{\omega^j} - \frac{1}{r}\sum_k \omega^k(\sum_t e_t\omega^t)\partial_{\omega^k} \\
&= \underline{\omega}\partial_r - \frac{\underline{\omega}}{r}\underline{\omega}\sum_j e^j\partial_{\omega^j} - \frac{\underline{\omega}}{r}\sum_k \omega^k\partial_{\omega^k}.
\end{aligned}$$

Next, keeping in mind the expression of the Dirac operator in spherical coordinates, the following expression for the angular Dirac operator Γ is obtained:

$$\Gamma = -\underline{\omega} \sum_j e^j \partial_{\omega^j} - \sum_k \omega^k \partial_{\omega^k}.$$

Consequently

$$\dot{\Gamma}\underline{\omega}\dot{f} = -\underline{\omega} \sum_k e^k \dot{\partial}_{\omega^k} \underline{\omega} \dot{f} - \sum_k \omega^k \dot{\partial}_{\omega^k} \underline{\omega} \dot{f} = -\underline{\omega} \sum_k e^k \underline{\omega} \frac{\partial f}{\partial \omega^k} - \underline{\omega} \sum_k \omega^k \frac{\partial f}{\partial \omega^k}.$$

From

$$e^k \underline{\omega} + \underline{\omega} e^k = e^k \sum_j \omega^j e_j + \sum_j \omega^j e_j e^k = \sum_j \omega^j (e^k e_j + e_j e^k) = -2\omega^k,$$

we infer

$$\begin{aligned}\dot{\Gamma}\underline{\omega}\dot{f} &= -\underline{\omega} \sum_k (-\underline{\omega} e^k - 2\omega^k) \frac{\partial f}{\partial \omega^k} - \underline{\omega} \sum_k \omega^k \frac{\partial f}{\partial \omega^k} = \underline{\omega}\underline{\omega} \sum_k e^k \frac{\partial f}{\partial \omega^k} + \underline{\omega} \sum_k \omega^k \frac{\partial f}{\partial \omega^k} \\ &= \underline{\omega} \left(\underline{\omega} \sum_k e^k \frac{\partial}{\partial \omega^k} + \sum_k \omega^k \frac{\partial}{\partial \omega^k} \right) f = -\underline{\omega} \Gamma f.\end{aligned}$$

Hence, the following commutation relation holds

$$\Gamma \underline{\omega} = (m-1)\underline{\omega} - \underline{\omega}\Gamma.$$

(ii) As $[\Gamma, r] = 0$, the above result (i) also yields

$$\Gamma \underline{x} = \underline{x}(m - 1 - \Gamma).$$

(iii) We now easily obtain

$$\begin{aligned}\Gamma \partial_{\underline{x}} = \Gamma \underline{\omega} \left(\partial_r + \frac{1}{r}\Gamma \right) &= \underline{\omega}(m-1-\Gamma) \left(\partial_r + \frac{1}{r}\Gamma \right) \\ &= \underline{\omega} \left(\partial_r + \frac{1}{r}\Gamma \right) (m-1-\Gamma) = \partial_{\underline{x}}(m-1-\Gamma). \qquad \square\end{aligned}$$

2.6. Solid g-spherical harmonics

In Subsection 2.4 we have introduced the g-Laplacian

$$\Delta_g = g^{jk} \partial^2_{x^j x^k}$$

and shown that it is factorized by the Dirac operator:

$$\partial^2_{\underline{x}} = -\Delta_g.$$

This leads inevitably to a relationship between the concepts of harmonicity and monogenicity.

Definition 2.44. A g-harmonic homogeneous polynomial S_k of degree k in $\mathbb{R}^m$:

$$S_k(\underline{x}) \in \mathcal{P}_k \quad \text{and} \quad \Delta_g[S_k(\underline{x})] = 0,$$

is called a *solid g-spherical harmonic.*

The space of solid g-spherical harmonics is denoted by $H(k)$.

Obviously we have that

$$M_\ell^+(k) \subset H(k) \quad \text{and} \quad M_r^+(k) \subset H(k).$$

Note also that for $P_{k-1} \in M_\ell^+(k-1)$

$$\underline{x} P_{k-1} \in H(k),$$

since, by Theorem 2.38,

$$\Delta_g[\underline{x} P_{k-1}] = (-\partial_{\underline{x}}^2)[\underline{x} P_{k-1}] = \partial_{\underline{x}}[(2k-2+m)P_{k-1}] = 0.$$

The following decomposition nicely illustrates that Clifford analysis is a refinement of harmonic analysis.

Proposition 2.45. *The space of solid g-spherical harmonics can be decomposed as*

$$H(k) = M_\ell^+(k) \oplus \underline{x} M_\ell^+(k-1).$$

Proof. For each $S_k \in H(k)$ we put

$$P_{k-1} = \partial_{\underline{x}}[S_k].$$

Naturally $P_{k-1} \in M_\ell^+(k-1)$, and by Theorem 2.38,

$$\partial_{\underline{x}}[\underline{x} P_{k-1}] = -(m+2k-2)P_{k-1}.$$

Hence

$$R_k = S_k + \frac{1}{m+2k-2}\, \underline{x} P_{k-1}$$

is left monogenic, which proves the statement. □

This monogenic decomposition now allows for an easy proof of the, coarser, g-harmonic decomposition of homogeneous polynomials.

Theorem 2.46 (g-harmonic (Fischer) decomposition). *Any $R_k \in \mathcal{P}_k$ has a unique decomposition of the form*

$$R_k(\underline{x}) = \sum_{2s \leq k} \|\underline{x}\|^{2s}\, S_{k-2s}(\underline{x}),$$

where the S_{k-2s} are solid g-spherical harmonics.

Proof. Take $R_{k-2} \in \mathcal{P}_{k-2}$ and $S_k \in H(k)$.

As $-\Delta_g$ is the Fischer dual of $\underline{x}^2$, we have

$$\begin{aligned}(\underline{x}^2 R_{k-2}(\underline{x}), S_k(\underline{x})) &= -\left[R_{k-2}^\dagger\left(g^{1j}\partial_{x^j}, \ldots, g^{mj}\partial_{x^j}\right)\Delta_g[S_k(\underline{x})]\right]_0 \\ &= -\left(R_{k-2}(\underline{x}), \Delta_g[S_k(\underline{x})]\right).\end{aligned}$$

Similarly as in the proof of Theorem 2.24, this result implies that any $R_k \in \mathcal{P}_k$ may be uniquely decomposed as

$$R_k = S_k + \|\underline{x}\|^2 R_{k-2}; \quad S_k \in H(k), \quad R_{k-2} \in \mathcal{P}_{k-2}.$$

Recursive application of the above result then indeed yields:

$$R_k(\underline{x}) = \sum_{2s \leq k} \|\underline{x}\|^{2s} S_{k-2s}(\underline{x}) \quad \text{with} \quad S_{k-2s} \in H(k-2s). \qquad \square$$

We end this subsection with the introduction of the so-called *g-Laplace-Beltrami-operator.*

We start from the decomposition of the g-Laplacian established in Corollary 2.33 and pass on to spherical coordinates:

$$\begin{aligned}
\Delta_g &= (r\partial_r + m - \Gamma) \frac{1}{r^2} (r\partial_r + \Gamma) \\
&= r\partial_r \frac{1}{r^2}(r\partial_r + \Gamma) + \frac{m}{r^2} (r\partial_r + \Gamma) - \Gamma \frac{1}{r^2} (r\partial_r + \Gamma) \\
&= r(-2)\frac{1}{r^3}(r\partial_r + \Gamma) + \frac{1}{r}\partial_r(r\partial_r + \Gamma) + \frac{m}{r}\partial_r + \frac{m}{r^2}\Gamma - \Gamma\frac{1}{r}\partial_r - \frac{1}{r^2}\Gamma^2 \\
&= -\frac{2}{r}\partial_r - \frac{2}{r^2}\Gamma + \frac{1}{r}\partial_r + \partial_r^2 + \frac{1}{r}\partial_r\Gamma + \frac{m}{r}\partial_r + \frac{m}{r^2}\Gamma - \frac{1}{r}\partial_r\Gamma - \frac{1}{r^2}\Gamma^2 \\
&= \partial_r^2 + \frac{(m-1)}{r}\partial_r + \frac{1}{r^2}[(m-2)\Gamma - \Gamma^2]
\end{aligned}$$

or finally

$$\Delta_g = \partial_r^2 + \frac{m-1}{r}\partial_r + \frac{1}{r^2}\Delta_g^*,$$

where we have put

$$\Delta_g^* = ((m-2) - \Gamma)\Gamma.$$

The operator Δ_g^* is called the g-Laplace-Beltrami operator. Note that this operator commutes with the angular Dirac operator:

$$[\Gamma, \Delta_g^*] = 0.$$

It follows that also the g-Laplace operator commutes with the angular Dirac operator:

$$[\Gamma, \Delta_g] = 0.$$

2.7. The g-Fourier transform

Let h be a positive function on $\mathbb{R}^m$. Then we consider the Clifford algebra-valued inner product

$$\langle f, g\rangle = \int_{\mathbb{R}^m} h(\underline{x}) \, f^\dagger(\underline{x}) \, g(\underline{x}) \, dx^M$$

where dx^M stands for the Lebesgue measure on $\mathbb{R}^m$; the associated norm then reads

$$\|f\|^2 = [\langle f, f\rangle]_0.$$

The unitary right Clifford-module of Clifford algebra-valued measurable functions on $\mathbb{R}^m$ for which $\|f\|^2 < \infty$ is a right Hilbert-Clifford-module which we denote by $L_2(\mathbb{R}^m, h(\underline{x}) \, dx^M)$.

In particular, taking $h(\underline{x}) \equiv 1$, we obtain the right Hilbert-module of *square integrable functions*:

$$L_2\left(\mathbb{R}^m, dx^M\right) = \left\{ f : \text{Lebesgue measurable in } \mathbb{R}^m \text{ for which} \right.$$
$$\left. \|f\|_2 = \left(\int_{\mathbb{R}^m} |f(\underline{x})|_0^2 \, dx^M \right)^{1/2} < \infty \right\}.$$

Naturally, $L_1\left(\mathbb{R}^m, dx^M\right)$ denotes the right Clifford-module of *integrable functions*:

$$L_1\left(\mathbb{R}^m, dx^M\right) = \left\{ f : \text{Lebesgue measurable in } \mathbb{R}^m \text{ for which} \right.$$
$$\left. \|f\|_1 = \int_{\mathbb{R}^m} |f(\underline{x})|_0 \, dx^M < \infty \right\}.$$

Now we define a *g-Fourier transform* on $L_1(\mathbb{R}^m, dx^M)$-functions, which by a classical argument may be extended to $L_2(\mathbb{R}^m, dx^M)$.

This g-Fourier transform of f, denoted by $\mathcal{F}_g[f]$, is defined by:

$$\begin{aligned}
\mathcal{F}_g[f](\underline{y}) &= \left(\frac{1}{\sqrt{2\pi}}\right)^m \int_{\mathbb{R}^m} \exp\left(-i\langle \underline{x}, \underline{y}\rangle\right) f(\underline{x}) \, dx^M \\
&= \left(\frac{1}{\sqrt{2\pi}}\right)^m \int_{\mathbb{R}^m} \exp\left(-i g_{jk} x^j y^k\right) f(x^1, \ldots, x^m) \, dx^M.
\end{aligned}$$

To show that this definition is meaningful in this metric dependent context, it is immediately checked how this g-Fourier transform behaves with respect to multiplication with the variable $\underline{x}$ and, by duality, with respect to the action of the Dirac operator.

Proposition 2.47. *The g-Fourier transform $\mathcal{F}_g$ satisfies:*

(i) *the multiplication rule:* $\mathcal{F}_g[\underline{x} f(\underline{x})](\underline{y}) = i\partial_{\underline{y}} \, \mathcal{F}_g[f(\underline{x})](\underline{y})$;

(ii) *the differentiation rule:* $\mathcal{F}_g[\partial_{\underline{x}} f(\underline{x})](\underline{y}) = i\underline{y} \, \mathcal{F}_g[f(\underline{x})](\underline{y})$.

Proof. We have consecutively

$$\begin{aligned}
\text{(i)} \quad \mathcal{F}_g[\underline{x} f(\underline{x})](\underline{y}) &= \sum_j e_j \left(\frac{1}{\sqrt{2\pi}}\right)^m \int_{\mathbb{R}^m} \exp\left(-i\langle \underline{x}, \underline{y}\rangle\right) x^j \, f(\underline{x}) \, dx^M \\
&= \sum_{j,k} g_{jk} e^k \left(\frac{1}{\sqrt{2\pi}}\right)^m \int_{\mathbb{R}^m} \exp\left(-i\langle \underline{x}, \underline{y}\rangle\right) x^j \, f(\underline{x}) \, dx^M \\
&= \sum_k e^k \left(\frac{1}{\sqrt{2\pi}}\right)^m \int_{\mathbb{R}^m} \exp\left(-i\langle \underline{x}, \underline{y}\rangle\right) \sum_j g_{jk} x^j \, f(\underline{x}) \, dx^M \\
&= \sum_k e^k \left(\frac{1}{\sqrt{2\pi}}\right)^m \int_{\mathbb{R}^m} i\partial_{y^k} \left[\exp\left(-i\langle \underline{x}, \underline{y}\rangle\right)\right] \, f(\underline{x}) \, dx^M \\
&= i\partial_{\underline{y}} \, \mathcal{F}_g[f(\underline{x})](\underline{y}).
\end{aligned}$$

$$\text{(ii)} \quad \mathcal{F}_g[\partial_{\underline{x}} f(\underline{x})](\underline{y}) = \sum_j e^j \left(\frac{1}{\sqrt{2\pi}}\right)^m \int_{\mathbb{R}^m} \exp\left(-i\langle \underline{x}, \underline{y}\rangle\right) \partial_{x^j}[f(\underline{x})]\, dx^M$$

$$= \sum_j e^j \left(\frac{1}{\sqrt{2\pi}}\right)^m \left\{ \int_{\mathbb{R}^m} \partial_{x^j} \Big[\exp\left(-i\langle \underline{x}, \underline{y}\rangle\right) f(\underline{x})\Big]\, dx^M \right.$$

$$\left. - \int_{\mathbb{R}^m} \partial_{x^j} \left[\exp\left(-i\langle \underline{x}, \underline{y}\rangle\right)\right] f(\underline{x})\, dx^M \right\}$$

$$= i \sum_j e^j \left(\frac{1}{\sqrt{2\pi}}\right)^m \int_{\mathbb{R}^m} \exp\left(-i\langle \underline{x}, \underline{y}\rangle\right) \sum_k g_{jk} y^k\, f(\underline{x})\, dx^M$$

$$= i \left(\frac{1}{\sqrt{2\pi}}\right)^m \int_{\mathbb{R}^m} \exp\left(-i\langle \underline{x}, \underline{y}\rangle\right) \sum_k e_k y^k\, f(\underline{x})\, dx^M$$

$$= i\underline{y}\; \mathcal{F}_g[f(\underline{x})](\underline{y}). \qquad \square$$

Another interesting question which is raised at once is a possible relationship with the classical Fourier transform $\mathcal{F}$ given by

$$\mathcal{F}[f](\underline{y}) = \left(\frac{1}{\sqrt{2\pi}}\right)^m \int_{\mathbb{R}^m} \exp\left(-i\sum_j x^j y^j\right) f(x^1, \ldots, x^m)\, dx^M.$$

To this end we will use again the basis $(E_j \;:\; j = 1, \ldots, m)$ consisting of eigenvectors of the matrix $G = (g_{jk})$ and the orthonormal frame $(\underline{\eta}_j \;:\; j = 1, \ldots, m)$ (see Subsection 2.3).

Proposition 2.48. *One has*

$$\mathcal{F}_g[f(x^j)](y^j) = \frac{1}{\sqrt{\lambda_1 \ldots \lambda_m}}\, \mathcal{F}[f(AP^{-1}(x'^j))](PA^T(y^j)).$$

Proof. By definition we have

$$\mathcal{F}_g[f(x^j)](y^j) = \left(\frac{1}{\sqrt{2\pi}}\right)^m \int_{\mathbb{R}^m} \exp\left(-i((x^j)^T G(y^j))\right) f(x^j)\, dx^M$$

$$= \left(\frac{1}{\sqrt{2\pi}}\right)^m \int_{\mathbb{R}^m} \exp\left(-i((x^j)^T GAP^{-1}(y'^j))\right) f(x^j)\, dx^M.$$

By means of the substitution $(x^j) = AP^{-1}(x'^j)$ for which

$$dx^M = \frac{1}{\sqrt{\lambda_1 \ldots \lambda_m}}\, dx'^M,$$

this becomes

$$\mathcal{F}_g[f(x^j)](y^j) = \frac{1}{\sqrt{\lambda_1 \ldots \lambda_m}} \left(\frac{1}{\sqrt{2\pi}}\right)^m \int_{\mathbb{R}^m} \exp\left(-i(x'^j)^T(y'^j)\right) f(AP^{-1}(x'^j))\, dx'^M$$

$$= \frac{1}{\sqrt{\lambda_1 \ldots \lambda_m}}\, \mathcal{F}[f(AP^{-1}(x'^j))](PA^T(y^j)). \qquad \square$$

This result now allows us to prove a Parseval formula for the g-Fourier transform.

Theorem 2.49. *The g-Fourier transform $\mathcal{F}_g$ satisfies the Parseval formula*

$$\langle f, h\rangle \; = \lambda_1 \ldots \lambda_m \; \langle \mathcal{F}_g[f], \mathcal{F}_g[h]\rangle, \quad f, h \in L_2(\mathbb{R}^m, dx^M). \tag{2.9}$$

Proof. Applying Proposition 2.48 results into

$$\begin{aligned}\langle \mathcal{F}_g[f(x^j)], \mathcal{F}_g[h(x^j)]\rangle &= \int_{\mathbb{R}^m} \left(\mathcal{F}_g[f(x^j)](y^j)\right)^\dagger \; \mathcal{F}_g[h(x^j)](y^j) \; dy^1 \ldots dy^m \\ &= \frac{1}{\lambda_1 \ldots \lambda_m} \int_{\mathbb{R}^m} \left(\mathcal{F}[f(AP^{-1}(x'^j))](PA^T(y^j))\right)^\dagger \\ &\qquad \mathcal{F}[h(AP^{-1}(x'^j))](PA^T(y^j)) \; dy^1 \ldots dy^m.\end{aligned}$$

By means of the substitution $(z^j) = PA^T(y^j)$ or equivalently $(y^j) = AP^{-1}(z^j)$ for which

$$dy^1 \ldots dy^m = \frac{1}{\sqrt{\lambda_1 \ldots \lambda_m}} \; dz^1 \ldots dz^m,$$

this becomes

$$\begin{aligned}\langle \mathcal{F}_g[f(x^j)], \mathcal{F}_g[h(x^j)]\rangle &= \frac{1}{\lambda_1 \ldots \lambda_m} \frac{1}{\sqrt{\lambda_1 \ldots \lambda_m}} \int_{\mathbb{R}^m} \left(\mathcal{F}[f(AP^{-1}(x'^j))](z^j)\right)^\dagger \\ &\qquad \mathcal{F}[h(AP^{-1}(x'^j))](z^j) \; dz^1 \ldots dz^m.\end{aligned}$$

Next, applying the Parseval formula for the classical Fourier transform $\mathcal{F}$ yields

$$\begin{aligned}&\langle \mathcal{F}_g[f(x^j)], \mathcal{F}_g[h(x^j)]\rangle \\ &= \frac{1}{\lambda_1 \ldots \lambda_m} \frac{1}{\sqrt{\lambda_1 \ldots \lambda_m}} \int_{\mathbb{R}^m} f^\dagger(AP^{-1}(x'^j)) \; h(AP^{-1}(x'^j)) \; dx'^1 \ldots dx'^m.\end{aligned}$$

Finally, the substitution $(u^j) = AP^{-1}(x'^j)$ or equivalently $(x'^j) = PA^T(u^j)$ for which

$$dx'^1 \ldots dx'^m = \sqrt{\lambda_1 \ldots \lambda_m} \; du^1 \ldots du^m$$

leads to the desired result:

$$\begin{aligned}\langle \mathcal{F}_g[f(x^j)], \mathcal{F}_g[h(x^j)]\rangle &= \frac{1}{\lambda_1 \ldots \lambda_m} \int_{\mathbb{R}^m} f^\dagger(u^j) \; h(u^j) \; du^1 \ldots du^m \\ &= \frac{1}{\lambda_1 \ldots \lambda_m} \langle f, h\rangle.\end{aligned}$$

□

Remark 2.50. Note that the classical Fourier transform and the g-Fourier transform are also related by:

$$\mathcal{F}_g[f(x^j)](y^j) = \mathcal{F}[f(x^j)](G(y^j))$$

allowing for a shorter proof of the Parseval formula (2.9).

Indeed, by means of the substitution $(u^j) = G(y^j)$ or equivalently $(y^j) = G^{-1}(u^j)$ for which

$$dy^M = \frac{du^M}{|\det(G)|} = \frac{1}{\lambda_1 \ldots \lambda_m} \; du^M,$$

we find

$$\begin{aligned}\langle \mathcal{F}_g[f(x^j)], \mathcal{F}_g[h(x^j)]\rangle &= \int_{\mathbb{R}^m} \left(\mathcal{F}[f(x^j)](G(y^j))\right)^\dagger \; \mathcal{F}[h(x^j)](G(y^j))\; dy^M \\ &= \frac{1}{\lambda_1 \dots \lambda_m} \int_{\mathbb{R}^m} \left(\mathcal{F}[f(x^j)](u^j)\right)^\dagger \; \mathcal{F}[h(x^j)](u^j)\; du^M \\ &= \frac{1}{\lambda_1 \dots \lambda_m} \langle f, h\rangle .\end{aligned}$$

3. Metric invariant integration theory

3.1. The basic language of Clifford differential forms

In this subsection we gather some basic definitions and properties concerning Clifford differential forms in the metric dependent setting.

Next to the contravariant vector of *coordinates* $(x^1, \dots, x^m)$ we consider the contravariant tensor of *basic differentials* $(dx^1, \dots, dx^m)$.

By means of the defining relations:

$$\begin{aligned} dx^j dx^j &= 0, \quad j = 1, \dots, m \\ dx^j dx^k &= -dx^k dx^j, \quad j \neq k, \quad j,k = 1, \dots, m, \end{aligned}$$

these basic differentials generate the Grassmann algebra or exterior algebra over $\mathbb{R}^m$

$$\Lambda \mathbb{R}^m = \sum_{r=0}^{m} \oplus \Lambda^r \mathbb{R}^m ,$$

where $\Lambda^r\mathbb{R}^m$ denotes the space of real-valued r-forms.

The basic toolkit for Clifford analysis is now extended to the "algebra of Clifford forms"

$$\Phi = \mathrm{Alg}\{x^1, \dots, x^m; dx^1, \dots, dx^m; e_1, \dots, e_m; i\}.$$

A Clifford form $F \in \Phi$ thus takes the form

$$F = \sum_A F_A \; dx^A$$

with

$$F_A = \sum_B f_{A,B}(\underline{x})\; e_B \quad , \quad f_{A,B} : \mathbb{R}^m \to \mathbb{C}$$

and

$$dx^A = dx^{i_1} dx^{i_2} \dots dx^{i_r} \quad , \quad A = (i_1, i_2, \dots, i_r).$$

Equivalently, we can also write

$$F = \sum_B G_B \; e_B$$

with

$$G_B = \sum_A f_{A,B}(\underline{x})\; dx^A .$$

The *exterior derivative* d is defined by means of the following axioms: for a C_1-form $F \in \Phi$ one has

(A1) $d(x^j F) = dx^j F + x^j\ dF$
(A2) $d(dx^j F) = -dx^j dF$
(A3) $d(e_j F) = e_j\ dF.$

We also have the property $d^2 = 0$ and with respect to the local coordinate system, d may be expressed as:

$$d = \partial_{x^j}\ dx^j .$$

Hence for a C_1 form $F = \sum_A F_A\ dx^A$ we have

$$dF = \sum_A \sum_j \partial_{x^j} [F_A]\ dx^j dx^A .$$

We also mention the following basic result, perhaps the most fundamental and most intriguing theorem in mathematical analysis.

Theorem 3.1 (Stokes's theorem). *Let Σ be an oriented k-surface in $\mathbb{R}^m$ and F a continuously differentiable $(k-1)$-form. Then one has:*

$$\int_{\partial\Sigma} F = \int_\Sigma dF.$$

Next we define the *basic contraction operators*, determined by the relations

(C1) $\partial_{x^j} \rfloor (x^k F) = x^k\ \partial_{x^j} \rfloor F$
(C2) $\partial_{x^j} \rfloor (dx^k F) = \delta_j^k\ F - dx^k \partial_{x^j} \rfloor F$.

Note that, due to (C2), the contraction operators $\partial_{x^j} \rfloor$ are in fact a kind of derivative with respect to the generators dx^j. They are sometimes called *fermionic derivatives.*

The Dirac contraction operator is given by

$$\partial_{\underline{x}} \rfloor = e^j \partial_{x^j} \rfloor .$$

Before discussing the notion of directional derivative and Lie derivative, we must point out a subtle difference between *vector fields* and *Clifford vector fields.* In differential geometry an operator of the form

$$v = v^j(\underline{x})\ \partial_{x^j} \quad , \quad v^j\ :\ \mathbb{R}^m \to \mathbb{C} \tag{3.1}$$

is called a "vector field with components v^j", whereas a Clifford vector field has the form

$$\underline{v} = v^j(\underline{x})\ e_j \quad , \quad v^j\ :\ \mathbb{R}^m \to \mathbb{C}.$$

For a survey paper on differential forms versus multi-vector functions, see [3].

Moreover, in Clifford analysis one may even consider vector fields (3.1) with $\mathbb{C}_m$-valued components v^j, which leads to several new interesting possibilities.

First of all, given a Clifford vector field

$$v = v^j(\underline{x})\ \partial_{x^j} \quad , \quad v^j\ :\ \mathbb{R}^m \to \mathbb{C}_m,$$

one may consider for any differential form $F = \sum_A F_A \, dx^A \in \Phi$, the *directional derivative* (or Levi-Civita connection)

$$v[F] = v^j(\underline{x}) \, \partial_{x^j}[F] = \sum_A v[F_A] \, dx^A$$

and contraction

$$v\rfloor \, F = v^j \, \partial_{x^j} \rfloor F.$$

One may also consider the *Lie derivative in the direction of* v

$$\mathcal{L}_v F = (v\rfloor \, d + d \, v\rfloor)F$$

that satisfies

$$\mathcal{L}_v \, d = d \, \mathcal{L}_v.$$

The *Leray form* $L(\underline{x}, d\underline{x})$ is defined as

$$L(\underline{x}, d\underline{x}) = E\rfloor dx^M = \sum_j (-1)^{j+1} x^j dx^{M\setminus\{j\}}$$

with

$$E\rfloor = x^j \partial_{x^j} \rfloor \qquad \text{and} \qquad dx^M = dx^1 \ldots dx^m$$

and finally

$$dx^{M\setminus\{j\}} = dx^1 \ldots [dx^j] \ldots dx^m.$$

This Leray form $L(\underline{x}, d\underline{x})$, which we will use in the sequel, is naturally closely related to the elementary volume element form dx^M. We have two formulae as to that issue.

Proposition 3.2. *One has*

(i) $$dL(\underline{x}, d\underline{x}) = m \, dx^M;$$

(ii) $$dx^M = \frac{1}{r} \, dr \, L(\underline{x}, d\underline{x}).$$

Proof. (i) In a straightforward way we obtain

$$dL(\underline{x}, d\underline{x}) = \sum_{j=1}^m (-1)^{j+1} \, dx^j dx^{M\setminus\{j\}} = \sum_{j=1}^m dx^M = m \, dx^M.$$

(ii) Differentiating the relation

$$r^2 = \sum_{j,k} g_{jk} x^j x^k = \sum_j g_{jj} (x^j)^2 + 2 \sum_{j<k} g_{jk} x^j x^k,$$

yields

$$2r \, dr = 2 \sum_j g_{jj} \, x^j dx^j + 2 \sum_{j<k} g_{jk} \, dx^j x^k + 2 \sum_{j<k} g_{jk} \, x^j dx^k.$$

Hence we have

$$
\begin{aligned}
& r\ dr\ L(\underline{x}, d\underline{x}) \\
&= \sum_j g_{jj}\ x^j\ dx^j \left(\sum_i (-1)^{i+1} x^i dx^{M\setminus\{i\}} \right) \\
&\quad + \sum_{j<k} g_{jk}\ x^k\ dx^j \left(\sum_i (-1)^{i+1} x^i\ dx^{M\setminus\{i\}} \right) \\
&\quad + \sum_{j<k} g_{jk}\ x^j\ dx^k \left(\sum_i (-1)^{i+1} x^i dx^{M\setminus\{i\}} \right) \\
&= \sum_j g_{jj}\ (x^j)^2\ (-1)^{j+1} dx^j dx^{M\setminus\{j\}} + \sum_{j<k} g_{jk}\ x^k x^j\ (-1)^{j+1}\ dx^j dx^{M\setminus\{j\}} \\
&\quad + \sum_{j<k} g_{jk}\ x^j x^k\ (-1)^{k+1}\ dx^k dx^{M\setminus\{k\}} \\
&= \sum_j g_{jj}\ (x^j)^2 dx^M + \sum_{j<k} g_{jk}\ x^k x^j dx^M + \sum_{j<k} g_{jk}\ x^j x^k dx^M \\
&= \left(\sum_j g_{jj}\ (x^j)^2 + 2 \sum_{j<k} g_{jk}\ x^j x^k \right)\ dx^M \\
&= r^2\ dx^M,
\end{aligned}
$$

which proves the statement. □

It is clear that the Leray form is homogeneous of degree m, i.e., it transforms like

$$L\big(\lambda\underline{x}, d(\lambda\underline{x})\big) = L(\lambda\underline{x}, \lambda d\underline{x} + \underline{x} d\lambda) = \lambda^m\ L(\underline{x}, d\underline{x}).$$

Hence, we define the associated Leray form of degree 0:

$$L(\underline{\omega}, d\underline{\omega}) := \frac{L(\underline{x}, d\underline{x})}{r^m}, \quad \underline{\omega} \in S^{m-1},$$

so that, by Proposition 3.2

$$dx^M = r^{m-1}\ dr\ L(\underline{\omega}, d\underline{\omega}).$$

The last differential form needed, is the so-called *sigma form* $\underline{\sigma}(\underline{x}, d\underline{x})$, defined as

$$\underline{\sigma}(\underline{x}, d\underline{x}) = \partial_{\underline{x}} \rfloor\ dx^M = \sum_j (-1)^{j+1} e^j dx^{M\setminus\{j\}}$$

and we shall see that it can be interpreted as the oriented surface element on a hypersurface.

A first auxiliary property reads as follows.

Lemma 3.3. *One has*

$$d\underline{\sigma}(\underline{x}, d\underline{x}) = 0.$$

Proof. As $d^2 = 0$, we immediately have

$$d\underline{\sigma}(\underline{x}, d\underline{x}) = \sum_j (-1)^{j+1} \; e^j \; d(dx^{M\setminus\{j\}}) = 0. \qquad \square$$

Next we search for a relationship between the Leray and the sigma form.

Theorem 3.4. *In the whole of* $\mathbb{R}^m$ *one has*

$$\underline{x} \; \underline{\sigma}(\underline{x}, d\underline{x}) + \underline{\sigma}(\underline{x}, d\underline{x}) \; \underline{x} = -2L(\underline{x}, d\underline{x}).$$

Proof. As

$$e_k e^j + e^j e_k = -2\delta_k^j,$$

we easily find by a direct computation

$$\begin{aligned}
&\underline{x} \; \underline{\sigma}(\underline{x}, d\underline{x}) + \underline{\sigma}(\underline{x}, d\underline{x}) \; \underline{x} \\
&= \sum_k x^k e_k \sum_j (-1)^{j+1} e^j dx^{M\setminus\{j\}} + \sum_j (-1)^{j+1} e^j dx^{M\setminus\{j\}} \sum_k x^k e_k \\
&= \sum_{j,k} x^k (-1)^{j+1} dx^{M\setminus\{j\}} (e_k e^j + e^j e_k) \\
&= -2 \sum_j x^j (-1)^{j+1} dx^{M\setminus\{j\}} = -2L(\underline{x}, d\underline{x}).
\end{aligned} \qquad \square$$

Remark 3.5. The relationship established in Theorem 3.4 may also be written as

$$\{\underline{x} \; , \; \underline{\sigma}(\underline{x}, d\underline{x})\} = -2L(\underline{x}, d\underline{x}) \qquad \text{or} \qquad \underline{x} \bullet \underline{\sigma}(\underline{x}, d\underline{x}) = -L(\underline{x}, d\underline{x}).$$

Finally we look for an expression of the sigma form in terms of the classical elementary hypersurface element. To that end consider a smooth hypersurface Σ in $\mathbb{R}^m$. Let $\underline{x}$ be an arbitrary point on Σ; then we call $\underline{n}(\underline{x})$ a unit Clifford vector along the surface normal to Σ at the point $\underline{x}$. By definition $\underline{n}(\underline{x})$ is orthogonal to the tangent space of Σ at $\underline{x}$, and so, putting $d\underline{x} = \sum_{j=1}^m e_j dx^j$:

$$\langle \underline{n}, d\underline{x} \rangle \; = 0$$

or

$$\sum_k \sum_\ell g_{k\ell} \; n^k dx^\ell = 0. \tag{3.2}$$

If, e.g., Σ is given by the cartesian equation $\varphi(\underline{x}) = 0$, then a Clifford normal vector at $\underline{x}$ is given by

$$\underline{N}(\underline{x}) = \sum_k e_k N^k = \sum_k e_k \sum_j \partial_{x^j}[\varphi] \; g^{jk} = \sum_k e_k \; \left(\vec{\nabla}\varphi\right) \; (g_{jk})^{-1}$$

for which indeed

$$\begin{aligned}
\langle \underline{N}(\underline{x}), d\underline{x} \rangle &= \sum_k \sum_\ell g_{k\ell} \; N^k dx^\ell = \sum_k \sum_\ell \sum_j g_{k\ell} \; \partial_{x^j}[\varphi] \; g^{jk} \; dx^\ell \\
&= \sum_\ell \sum_j \delta_\ell^j \; \partial_{x^j}[\varphi] \; dx^\ell = \sum_j \partial_{x^j}[\varphi] \; dx^j = d\varphi = 0.
\end{aligned}$$

Note that $\underline{N}(\underline{x})$ is not a unit vector since

$$\|\underline{N}(\underline{x})\|^2 = \langle \underline{N}(\underline{x}), \underline{N}(\underline{x}) \rangle = \sum_j \sum_k g_{jk}\ N^j N^k$$
$$= \sum_j \sum_k \sum_t \sum_s g_{jk}\ \partial_{x^t}[\varphi]\ g^{tj}\ \partial_{x^s}[\varphi]\ g^{sk}$$
$$= \sum_k \sum_t \sum_s \delta_k^t\ \partial_{x^t}[\varphi]\ \partial_{x^s}[\varphi]\ g^{sk} = \sum_t \sum_s \partial_{x^t}[\varphi]\ \partial_{x^s}[\varphi]\ g^{st}$$

and hence

$$\underline{n}(\underline{x}) = \frac{\sum_k \sum_j e_k\ \partial_{x^j}[\varphi]\ g^{jk}}{\left(\sum_k \sum_j \partial_{x^k}[\varphi]\ \partial_{x^j}[\varphi]\ g^{jk}\right)^{1/2}}.$$

Now it may be proved that the sigma form is also orthogonal to the tangent space or, equivalently, that the sigma form lies along the surface normal at the point considered.

Proposition 3.6. *Let Σ be a smooth hypersurface and $\underline{n}(\underline{x})$ a unit normal Clifford vector at $\underline{x} \in \Sigma$, then*

$$\underline{n}(\underline{x}) \wedge \underline{\sigma}(\underline{x}, d\underline{x}) = 0.$$

Proof. By definition we have

$$\underline{n}(\underline{x}) \wedge \underline{\sigma}(\underline{x}, d\underline{x}) = \left(\sum_k e_k n^k\right) \wedge \left(\sum_j (-1)^{j+1} e^j\, dx^{M\setminus\{j\}}\right)$$
$$= \left(\sum_{k,t} g_{kt}\ e^t n^k\right) \wedge \left(\sum_j (-1)^{j+1} e^j\, dx^{M\setminus\{j\}}\right)$$
$$= \sum_k n^k \sum_{t,j} (-1)^{j+1} g_{kt}\ e^t \wedge e^j\ dx^{M\setminus\{j\}}. \tag{3.3}$$

Note that the above expression is a Clifford bivector; furthermore, as $e^j \wedge e^j = 0$, only the terms with $t \neq j$ remain.

Take such a t and j fixed with, for example, $t < j$ and consider the coefficient of the bivector $e^t \wedge e^j$.

In view of the anti-commutativity of the outer product

$$e^t \wedge e^j = -e^j \wedge e^t,$$

this coefficient takes the form

$$\sum_k n^k\ (-1)^{j+1} g_{kt}\ dx^{M\setminus\{j\}} - \sum_k n^k\ (-1)^{t+1} g_{kj}\ dx^{M\setminus\{t\}}.$$

As, by assumption, $t < j$, the above expression can be rewritten as

$$\sum_k n^k \, (-1)^{j+1} \, g_{kt} \, (-1)^{t-1} \, dx^t dx^{M\backslash\{t,j\}}$$
$$- \sum_k n^k \, (-1)^{t+1} \, g_{kj} \, (-1)^{j-2} \, dx^j dx^{M\backslash\{t,j\}}$$
$$= (-1)^{j+t} \left(\sum_k n^k \, g_{kt} \, dx^t + \sum_k n^k \, g_{kj} \, dx^j \right) dx^{M\backslash\{t,j\}}.$$

Finally, by means of (3.2) we have

$$\left(\sum_{\ell,k} n^k g_{k\ell} \, dx^\ell \right) dx^{M\backslash\{t,j\}} = 0$$

or equivalently

$$\left(\sum_k n^k g_{kt} \, dx^t + \sum_k n^k g_{kj} \, dx^j \right) dx^{M\backslash\{t,j\}} = 0.$$

Hence, the coefficient of each bivector $e^t \wedge e^j$ in (3.3) is zero, which proves the statement. □

Corollary 3.7. *For any point $\underline{x}$ on the unit sphere S^{m-1} one has*

$$\underline{x} \wedge \underline{\sigma}(\underline{x}, d\underline{x}) = 0.$$

Proof. It suffices to note that for any point $\underline{x} \in S^{m-1}$, a unit normal vector is precisely given by $\underline{n}(\underline{x}) = \underline{x}$. □

Combining Theorem 3.4 and the previous Corollary 3.7 yields the following additional result on the unit sphere.

Corollary 3.8. *For each $\underline{x} \in S^{m-1}$, one has*

$$\underline{x} \; \underline{\sigma}(\underline{x}, d\underline{x}) = \underline{\sigma}(\underline{x}, d\underline{x}) \; \underline{x} = -L(\underline{x}, d\underline{x}) \quad \text{and} \quad \underline{\sigma}(\underline{x}, d\underline{x}) = \underline{x} \; L(\underline{x}, d\underline{x}) = L(\underline{x}, d\underline{x}) \; \underline{x}.$$

3.2. Orthogonal spherical monogenics

The aim of this section is to establish orthogonality relations between the inner and outer spherical monogenics on the unit sphere.

3.2.1. The Cauchy-Pompeiu formula.

Theorem 3.9 (Cauchy-Pompeiu). *Let Ω be a compact orientable m-dimensional manifold with boundary $\partial\Omega$ and $f, g \in C_1(\Omega)$. Then one has:*

$$\int_{\partial\Omega} f \; \underline{\sigma}(\underline{x}, d\underline{x}) \; g = \int_\Omega \left((f \partial_{\underline{x}}) g + f (\partial_{\underline{x}} g) \right) \, dx^M$$

and in particular, for $f \equiv 1$:

$$\int_{\partial\Omega} \underline{\sigma}(\underline{x}, d\underline{x}) \; g = \int_\Omega \partial_{\underline{x}}[g] \; dx^M.$$

Proof. Taking into account that the sigma form is an $(m-1)$ form, for which moreover $d\underline{\sigma}(\underline{x}, d\underline{x}) = 0$ (see Lemma 3.3), we have

$$\begin{aligned} d(f\ \underline{\sigma}(\underline{x}, d\underline{x})\ g) &= d\big(f\ \underline{\sigma}(\underline{x}, d\underline{x})\big)\ g + (-1)^{m-1} f\ \underline{\sigma}(\underline{x}, d\underline{x})\ dg \\ &= df\ \underline{\sigma}(\underline{x}, d\underline{x})\ g + (-1)^{m-1} f\ \underline{\sigma}(\underline{x}, d\underline{x})\ dg \\ &= \sum_j \partial_{x^j}[f]\ dx^j\ \underline{\sigma}(\underline{x}, d\underline{x})\ g \\ &\qquad + (-1)^{m-1}\ f\ \underline{\sigma}(\underline{x}, d\underline{x}) \Big(\sum_j dx^j\ \partial_{x^j}[g]\Big). \end{aligned} \tag{3.4}$$

Now for each $j = 1, 2, \ldots, m$ fixed, we obtain

$$\begin{aligned} dx^j\ \underline{\sigma}(\underline{x}, d\underline{x}) &= dx^j \Big(\sum_k (-1)^{k+1} e^k dx^{M\setminus\{k\}}\Big) = \sum_k (-1)^{k+1} e^k dx^j dx^{M\setminus\{k\}} \\ &= \sum_k (-1)^{k+1} e^k (-1)^{j-1} dx^M \delta_{j,k} = e^j dx^M \end{aligned}$$

and similarly

$$\begin{aligned} \underline{\sigma}(\underline{x}, d\underline{x})\ dx^j &= \sum_k (-1)^{k+1} e^k dx^{M\setminus\{k\}} dx^j = \sum_k (-1)^{k+1} e^k (-1)^{m-j} dx^M \delta_{j,k} \\ &= (-1)^{m+1} e^j dx^M. \end{aligned}$$

Hence, (3.4) becomes

$$\begin{aligned} d(f\ \underline{\sigma}(\underline{x}, d\underline{x})\ g) &= \sum_j \partial_{x^j}[f]\ e^j g\ dx^M + f \sum_j e^j \partial_{x^j}[g]\ dx^M \\ &= (f\partial_{\underline{x}}) g\ dx^M + f(\partial_{\underline{x}} g)\ dx^M. \end{aligned}$$

Consequently, by means of Stokes's theorem (Theorem 3.1) we indeed obtain

$$\begin{aligned} \int_{\partial\Omega} f\ \underline{\sigma}(\underline{x}, d\underline{x})\ g &= \int_\Omega d(f\ \underline{\sigma}(\underline{x}, d\underline{x})\ g) \\ &= \int_\Omega [(f\partial_{\underline{x}})g + f(\partial_{\underline{x}} g)]\ dx^M. \end{aligned}$$

□

The Cauchy-Pompeiu formula immediately yields the following fundamental result.

Corollary 3.10 (Cauchy's theorem). *Let Ω be a compact orientable m-dimensional manifold with boundary $\partial\Omega$. If f is right monogenic in Ω and g is left monogenic in Ω, one has:*

$$\int_{\partial\Omega} f\ \underline{\sigma}(\underline{x}, d\underline{x})\ g = 0.$$

3.2.2. Spherical monogenics.

Definition 3.11.

(i) The restriction to S^{m-1} of a solid inner spherical monogenic $P_k \in M_\ell^+(k)$ is called an *inner spherical monogenic*, while the restriction to S^{m-1} of a solid outer spherical monogenic $Q_k \in M_\ell^-(k)$ is called an *outer spherical monogenic*.

(ii) The restriction to S^{m-1} of a solid g-spherical harmonic $S_k \in H(k)$ is called a g-spherical harmonic.

The space of inner, respectively outer, spherical monogenics of order k is denoted by $\mathcal{M}_\ell^+(k)$, respectively $\mathcal{M}_\ell^-(k)$, while the space of g-spherical harmonics of degree k is denoted by $\mathcal{H}(k)$.

Inner and outer spherical monogenics on the unit sphere S^{m-1} are related as follows (see Proposition 2.39):

$$P_k(\underline{x}) \in \mathcal{M}_\ell^+(k) \iff \underline{x}P_k(\underline{x}) \in \mathcal{M}_\ell^-(k), \quad \underline{x} \in S^{m-1}.$$

This result combined with Corollary 3.8 enables us to prove the orthogonality of the spherical monogenics.

Theorem 3.12 (Orthogonality spherical monogenics).

(i) *Any inner and outer spherical monogenic are orthogonal, i.e., for all k and t one has*

$$\int_{S^{m-1}} Q_t^\dagger(\underline{x})\; P_k(\underline{x})\; L(\underline{x}, d\underline{x}) = 0.$$

(ii) *Inner spherical monogenics of different degree are orthogonal, i.e., for $t \neq k$ one has*

$$\int_{S^{m-1}} P_t^\dagger(\underline{x})\; P_k(\underline{x})\; L(\underline{x}, d\underline{x}) = 0.$$

(iii) *Outer spherical monogenics of different degree are orthogonal, i.e., for $t \neq k$ one has*

$$\int_{S^{m-1}} Q_t^\dagger(\underline{x})\; Q_k(\underline{x})\; L(\underline{x}, d\underline{x}) = 0.$$

Proof. The proof is based on Cauchy's theorem (Corollary 3.10) with Ω the closed unit ball

$$\Omega = B(1) = \{\underline{x} \in \mathbb{R}^1_m \;:\; \|\underline{x}\| = (g_{jk}x^j x^k)^{1/2} \leq 1\}$$

and $\partial\Omega$ the unit sphere:

$$\partial\Omega = S^{m-1} = \{\underline{x} \in \mathbb{R}^1_m \;:\; \underline{x}^2 = -\|\underline{x}\|^2 = -g_{jk}x^j x^k = -1\}.$$

(i) Take two arbitrary spherical monogenics $P_k \in \mathcal{M}_\ell^+(k)$ and $Q_t \in \mathcal{M}_\ell^-(t)$.

By means of Corollary 3.8 we have

$$\begin{aligned}\int_{S^{m-1}} Q_t^\dagger(\underline{x})\; P_k(\underline{x})\; L(\underline{x}, d\underline{x}) &= \int_{S^{m-1}} Q_t^\dagger(\underline{x})\; (-\underline{x}\; \underline{\sigma}(\underline{x}, d\underline{x}))\; P_k(\underline{x}) \\ &= \int_{S^{m-1}} (\underline{x}\; Q_t(\underline{x}))^\dagger\; \underline{\sigma}(\underline{x}, d\underline{x})\; P_k(\underline{x}). \qquad (3.5)\end{aligned}$$

As $Q_t \in \mathcal{M}_\ell^-(t)$, there exists $P_t \in M_\ell^+(t)$ such that

$$P_t(\underline{x}) = \underline{x}\ Q_t(\underline{x}), \quad \underline{x} \in S^{m-1}.$$

Hence equation (3.5) becomes

$$\int_{S^{m-1}} Q_t^\dagger(\underline{x})\ P_k(\underline{x})\ L(\underline{x}, d\underline{x}) = \int_{S^{m-1}} P_t^\dagger(\underline{x})\ \underline{\sigma}(\underline{x}, d\underline{x})\ P_k(\underline{x}).$$

Moreover, as $P_t^\dagger$ is right monogenic in $B(1)$, while P_k is left monogenic in $B(1)$, Cauchy's theorem yields

$$\int_{S^{m-1}} P_t^\dagger(\underline{x})\ \underline{\sigma}(\underline{x}, d\underline{x})\ P_k(\underline{x}) = 0.$$

(ii) Take $P_t \in \mathcal{M}_\ell^+(t)$ and $P_k \in \mathcal{M}_\ell^+(k)$ arbitrarily with $t \neq k$.

For the sake of clarity, we now use the notation $\underline{\omega}$ to denote an arbitrary element of the unit sphere S^{m-1}. Similar to (i), we have consecutively

$$\begin{aligned}\int_{S^{m-1}} P_t^\dagger(\underline{\omega})\ P_k(\underline{\omega})\ L(\underline{\omega}, d\underline{\omega}) &= \int_{S^{m-1}} (\underline{\omega} P_t(\underline{\omega}))^\dagger\ \underline{\sigma}(\underline{\omega}, d\underline{\omega})\ P_k(\underline{\omega}) \\ &= \int_{S^{m-1}} Q_t^\dagger(\underline{\omega})\ \underline{\sigma}(\underline{\omega}, d\underline{\omega})\ P_k(\underline{\omega})\end{aligned}$$

with $Q_t \in \mathcal{M}_\ell^-(t)$ such that

$$Q_t(\underline{\omega}) = \underline{\omega}\ P_t(\underline{\omega}).$$

By definition, there also exist $Q_t \in M_\ell^-(t)$ and $P_k \in M_\ell^+(k)$ such that

$$Q_t(\underline{x})/_{S^{m-1}} = Q_t(\underline{\omega}) \quad \text{and} \quad P_k(\underline{x})/_{S^{m-1}} = P_k(\underline{\omega}).$$

Next, consider the integral

$$\int_{\partial \overline{B}(\rho)} Q_t^\dagger(\underline{x})\ \underline{\sigma}(\underline{x}, d\underline{x})\ P_k(\underline{x}),$$

where $\overline{B}(\rho)$ denotes the closed ball with radius ρ

$$\overline{B}(\rho) = \{\underline{x} \in \mathbb{R}_m^1\ :\ \|\underline{x}\| = \left(g_{jk} x^j x^k\right)^{1/2} \leq \rho\}.$$

This integral is independent of the radius ρ. Indeed, applying Cauchy's theorem to the compact manifold $\Omega = \overline{B}(\rho) \setminus B(\widetilde{\rho})$ with boundary $\partial\Omega = \partial B(\rho) \cup \partial B(\widetilde{\rho})$ yields

$$\int_{\partial\Omega} Q_t^\dagger(\underline{x})\ \underline{\sigma}(\underline{x}, d\underline{x})\ P_k(\underline{x}) = 0$$

or, taking into account the orientation of $\underline{\sigma}(\underline{x}, d\underline{x})$,

$$\int_{\partial B(\rho)} Q_t^\dagger(\underline{x})\ \underline{\sigma}(\underline{x}, d\underline{x})\ P_k(\underline{x}) = \int_{\partial B(\widetilde{\rho})} Q_t^\dagger(\underline{x})\ \underline{\sigma}(\underline{x}, d\underline{x})\ P_k(\underline{x}).$$

Hence the following equality holds for all $\rho \in [0, \infty[$:

$$\int_{S^{m-1}} Q_t^\dagger(\underline{\omega})\ \underline{\sigma}(\underline{\omega}, d\underline{\omega})\ P_k(\underline{\omega}) = \int_{\partial \overline{B}(\rho)} Q_t^\dagger(\underline{x})\ \underline{\sigma}(\underline{x}, d\underline{x})\ P_k(\underline{x}).$$

Passing on to spherical coordinates $\underline{x} = \rho\underline{\omega}$, $\underline{\omega} \in S^{m-1}$ for all $\underline{x} \in \partial\overline{B}(\rho)$ gives

$$\int_{S^{m-1}} Q_t^\dagger(\underline{\omega})\ \underline{\sigma}(\underline{\omega}, d\underline{\omega})\ P_k(\underline{\omega}) = \int_{S^{m-1}} Q_t^\dagger(\rho\underline{\omega})\ \underline{\sigma}(\rho\underline{\omega}, d(\rho\underline{\omega}))\ P_k(\rho\underline{\omega})$$
$$= \rho^{k-t} \int_{S^{m-1}} Q_t^\dagger(\underline{\omega})\ \underline{\sigma}(\underline{\omega}, d\underline{\omega})\ P_k(\underline{\omega}).$$

Since, by assumption, $t \neq k$ the above equation implies

$$\int_{S^{m-1}} Q_t^\dagger(\underline{\omega})\ \underline{\sigma}(\underline{\omega}, d\underline{\omega})\ P_k(\underline{\omega}) = 0$$

which proves the statement.
(iii) Similar to (ii). □

4. The radial g-Clifford-Hermite polynomials and associated CCWT

4.1. The radial g-Clifford-Hermite polynomials

On the real line the Hermite polynomials may be defined by the Rodrigues formula

$$H_\ell(t) = (-1)^\ell \exp\left(\frac{t^2}{2}\right) \frac{d^\ell}{dt^\ell}\left(\exp\left(-\frac{t^2}{2}\right)\right), \qquad \ell = 0, 1, 2, \ldots$$

They constitute an orthogonal basis for the weighted Hilbert space

$$L_2\Big(\,]-\infty, +\infty[\ ,\ \exp\left(-\frac{t^2}{2}\right)\, dt\Big),$$

and satisfy the orthogonality relation

$$\int_{-\infty}^{+\infty} \exp\left(-\frac{t^2}{2}\right) H_\ell(t)\ H_k(t)\ dt = \ell!\ \sqrt{2\pi}\ \delta_{\ell,k}$$

and the recurrence relation

$$H_{\ell+1}(t) = t\ H_\ell(t) - \frac{d}{dt}\Big(H_\ell(t)\Big).$$

Furthermore, $H_\ell(t)$ is an even or an odd function according to the parity of ℓ, i.e., $H_\ell(-t) = (-1)^\ell H_\ell(t)$.

Similar to this classical situation and to the orthogonal Clifford analysis case (see [27]), the so-called *radial g-Clifford-Hermite polynomials* are defined by the Rodrigues formula

$$H_\ell(\underline{x}) = (-1)^\ell \exp\left(\frac{\|\underline{x}\|^2}{2}\right) \partial_{\underline{x}}^\ell \left[\exp\left(-\frac{\|\underline{x}\|^2}{2}\right)\right], \quad \ell = 0, 1, 2, \ldots. \tag{4.1}$$

They satisfy the following recurrence relation.

Proposition 4.1. *The radial g-Clifford-Hermite polynomials satisfy the recurrence relation*

$$H_{\ell+1}(\underline{x}) = (\underline{x} - \partial_{\underline{x}})[H_\ell(\underline{x})]. \tag{4.2}$$

Proof. As

$$\partial_{\underline{x}}\left[\exp\left(\frac{\|\underline{x}\|^2}{2}\right)\right] = \partial_{\underline{x}}\left[\exp\left(-\frac{\underline{x}^2}{2}\right)\right]$$
$$= \exp\left(-\frac{\underline{x}^2}{2}\right)\left(-\frac{1}{2}\right)(-2\underline{x}) = \underline{x}\exp\left(\frac{\|\underline{x}\|^2}{2}\right),$$

we have consecutively

$$\begin{aligned}\partial_{\underline{x}}[H_\ell(\underline{x})] &= (-1)^\ell\ \partial_{\underline{x}}\left[\exp\left(\frac{\|\underline{x}\|^2}{2}\right)\right]\ \partial_{\underline{x}}^\ell\left[\exp\left(-\frac{\|\underline{x}\|^2}{2}\right)\right]\\ &\quad + (-1)^\ell \exp\left(\frac{\|\underline{x}\|^2}{2}\right)\ \partial_{\underline{x}}^{\ell+1}\left[\exp\left(-\frac{\|\underline{x}\|^2}{2}\right)\right]\\ &= \underline{x}\ (-1)^\ell \exp\left(\frac{\|\underline{x}\|^2}{2}\right)\ \partial_{\underline{x}}^\ell\left[\exp\left(-\frac{\|\underline{x}\|^2}{2}\right)\right] - H_{\ell+1}(\underline{x})\\ &= \underline{x}\ H_\ell(\underline{x}) - H_{\ell+1}(\underline{x})\end{aligned}$$

which proves the statement. □

The recurrence relation (4.2) allows for a straightforward computation of some lower dimensional g-Clifford-Hermite polynomials:

$$\begin{aligned}H_0(\underline{x}) &= 1\\ H_1(\underline{x}) &= \underline{x}\\ H_2(\underline{x}) &= \underline{x}^2 + m = -\|\underline{x}\|^2 + m\\ H_3(\underline{x}) &= \underline{x}^3 + (m+2)\underline{x} = \underline{x}\ (-\|\underline{x}\|^2 + m + 2)\\ H_4(\underline{x}) &= \underline{x}^4 + 2(m+2)\underline{x}^2 + m(m+2) = \|\underline{x}\|^4 - 2(m+2)\ \|\underline{x}\|^2 + m(m+2)\\ &\text{etc.}\end{aligned}$$

Note that $H_\ell(\underline{x})$ is a polynomial of degree ℓ in the variable $\underline{x} = x^j e_j$, that $H_{2\ell}(\underline{x})$ only contains even powers of $\underline{x}$ and is scalar-valued, while $H_{2\ell+1}(\underline{x})$ only contains odd powers and is vector-valued.

By means of the Rodrigues formula (4.1) and the Cauchy-Pompeiu formula (Theorem 3.9) we obtain the following orthogonality relation.

Theorem 4.2. *The radial g-Clifford-Hermite polynomials $H_\ell(\underline{x})$ are mutually orthogonal in $\mathbb{R}^m$ with respect to the weight function* $\exp\left(-\frac{\|\underline{x}\|^2}{2}\right)$*, i.e., for $\ell \neq t$ one has*

$$\int_{\mathbb{R}^m} H_\ell^\dagger(\underline{x})\ H_t(\underline{x})\ \exp\left(-\frac{\|\underline{x}\|^2}{2}\right)\ dx^M = 0.$$

Proof. Suppose that $\ell < t$; the case $\ell > t$ follows by Hermitian conjugation.

As $H_\ell(\underline{x})$ is a polynomial of degree ℓ in $\underline{x}$, it is sufficient to show that for each $t \in \mathbb{N}$ and $\ell < t$:

$$\int_{\mathbb{R}^m} \underline{x}^\ell\ H_t(\underline{x})\ \exp\left(-\frac{\|\underline{x}\|^2}{2}\right)\ dx^M = 0.$$

We prove this by induction on ℓ. For $\ell = 0$ we have for each $t > 0$:

$$\begin{aligned}
&\int_{\mathbb{R}^m} H_t(\underline{x}) \, \exp\left(-\frac{\|\underline{x}\|^2}{2}\right) \, dx^M \\
&= (-1)^t \int_{\mathbb{R}^m} \partial_{\underline{x}}^t \left[\exp\left(-\frac{\|\underline{x}\|^2}{2}\right)\right] \, dx^M \\
&= (-1)^t \lim_{\rho\to\infty} \left(\int_{\partial \overline{B}(\rho)} \underline{\sigma}(\underline{x}, d\underline{x}) \, \partial_{\underline{x}}^{t-1} \left[\exp\left(-\frac{\|\underline{x}\|^2}{2}\right)\right]\right) = 0,
\end{aligned}$$

where we have used the Rodrigues formula (4.1) and the Cauchy-Pompeiu formula. Assume that orthogonality holds for $(\ell - 1)$ and $t > (\ell - 1)$. Take $t > \ell$. Again by means of the Rodrigues formula and the Cauchy-Pompeiu formula, we obtain:

$$\begin{aligned}
&\int_{\mathbb{R}^m} \underline{x}^\ell \, H_t(\underline{x}) \, \exp\left(-\frac{\|\underline{x}\|^2}{2}\right) \, dx^M \\
&= (-)^t \int_{\mathbb{R}^m} \underline{x}^\ell \, \partial_{\underline{x}}^t \left[\exp\left(-\frac{\|\underline{x}\|^2}{2}\right)\right] \, dx^M \\
&= (-1)^t \left\{ \lim_{\rho\to\infty} \int_{\partial \overline{B}(\rho)} \underline{x}^\ell \, \underline{\sigma}(\underline{x}, d\underline{x}) \, \partial_{\underline{x}}^{t-1} \left[\exp\left(-\frac{\|\underline{x}\|^2}{2}\right)\right] \right. \\
&\qquad \left. - \int_{\mathbb{R}^m} (\underline{x}^\ell \partial_{\underline{x}}) \, \partial_{\underline{x}}^{t-1} \left[\exp\left(-\frac{\|\underline{x}\|^2}{2}\right)\right] \, dx^M \right\} \\
&= (-1)^{t+1} \int_{\mathbb{R}^m} (\underline{x}^\ell \partial_{\underline{x}}) \, \partial_{\underline{x}}^{t-1} \left[\exp\left(-\frac{\|\underline{x}\|^2}{2}\right)\right] \, dx^M \\
&= \int_{\mathbb{R}^m} (\underline{x}^\ell \partial_{\underline{x}}) \, H_{t-1}(\underline{x}) \, \exp\left(-\frac{\|\underline{x}\|^2}{2}\right) \, dx^M.
\end{aligned}$$

From Theorem 2.38 we obtain in particular

$$\partial_{\underline{x}}[\underline{x}^\ell] = \begin{cases} -\ell \, \underline{x}^{\ell-1} & \text{for } \ell \text{ even} \\ -(\ell - 1 + m) \, \underline{x}^{\ell-1} & \text{for } \ell \text{ odd.} \end{cases}$$

Hence, by Hermitian conjugation we find

$$[\underline{x}^\ell]\partial_{\underline{x}} = \begin{cases} -\ell \, \underline{x}^{\ell-1} & \text{for } \ell \text{ even} \\ -(\ell - 1 + m) \, \underline{x}^{\ell-1} & \text{for } \ell \text{ odd.} \end{cases}$$

Summarizing: $[\underline{x}^\ell]\partial_{\underline{x}} \approx \underline{x}^{\ell-1}$, so that in view of the induction hypothesis

$$\int_{\mathbb{R}^m} \underline{x}^\ell \, H_t(\underline{x}) \, \exp\left(-\frac{\|\underline{x}\|^2}{2}\right) \, dx^M \approx \int_{\mathbb{R}^m} \underline{x}^{\ell-1} \, H_{t-1}(\underline{x}) \, \exp\left(-\frac{\|\underline{x}\|^2}{2}\right) \, dx^M = 0.$$

□

4.2 The g-Clifford-Hermite wavelets

As already explained in the introduction (Section 1) mother wavelets are L_2-functions, possibly with a number of vanishing moments, and satisfying an admissibility condition, turning the corresponding CWT into an isometry between the L_2-space of signals and the weighted L_2-space of transforms.

Now we show that the radial g-Clifford-Hermite polynomials constructed in the foregoing subsection are the building blocks for multi-dimensional Clifford mother wavelet functions.

For $t > 0$ Theorem 4.2 implies that

$$\int_{\mathbb{R}^m} H_t(\underline{x}) \exp\left(-\frac{\|\underline{x}\|^2}{2}\right) dx^M = 0.$$

In terms of wavelet theory this means that the alternatively scalar- or vector-valued $L_1 \cap L_2$-functions

$$\psi_t(\underline{x}) = H_t(\underline{x}) \exp\left(-\frac{\|\underline{x}\|^2}{2}\right) = (-1)^t\, \partial_{\underline{x}}^t \left[\exp\left(-\frac{\|\underline{x}\|^2}{2}\right)\right]$$

have zero momentum and are a good candidate for mother wavelets in $\mathbb{R}^m$, if at least they satisfy an appropriate admissibility condition (see Subsection 4.3). We call them the *g-Clifford-Hermite wavelets.*

The orthogonality relation of the radial g-Clifford-Hermite polynomials implies that the g-Clifford-Hermite wavelet ψ_t has vanishing moments up to order $(t-1)$:

$$\int_{\mathbb{R}^m} \underline{x}^\ell\, \psi_t(\underline{x})\, dx^M = 0, \quad \ell = 0, 1, \ldots, t-1.$$

As the capacity of wavelets for detecting singularities in a signal is related to their number of vanishing moments, this means that the g-Clifford-Hermite wavelet ψ_t is particularly appropriate for pointwise signal analysis whereby the corresponding CWT will filter out polynomial behavior of the signal up to degree $(t-1)$.

Next, we compute the g-Fourier transform of the g-Clifford-Hermite wavelets.

Lemma 4.3. *The g-Fourier transform of the g-Clifford-Hermite wavelets takes the form:*

$$\mathcal{F}_g[\psi_t(\underline{x})](\underline{y}) = \frac{1}{\sqrt{\lambda_1 \ldots \lambda_m}}\, (-i)^t\, \underline{y}^t\, \exp\left(-\frac{\|\underline{y}\|^2}{2}\right).$$

Proof. First, Proposition 2.47 yields

$$\mathcal{F}_g[\psi_t(\underline{x})](\underline{y}) = (-i)^t \underline{y}^t\, \mathcal{F}_g\left[\exp\left(-\frac{\|\underline{x}\|^2}{2}\right)\right](\underline{y}).$$

Moreover, by means of Proposition 2.48 and the fact that the function $\exp\left(-\frac{(x'^j)^T(x'^j)}{2}\right)$ is an eigenfunction of the classical Fourier transform:

$$\mathcal{F}\left[\exp\left(-\frac{(x'^j)^T(x'^j)}{2}\right)\right](y'^j) = \exp\left(-\frac{(y'^j)^T(y'^j)}{2}\right),$$

we obtain consecutively

$$\begin{aligned}
\mathcal{F}_g\left[\exp\left(-\frac{\|\underline{x}\|^2}{2}\right)\right](\underline{y}) &= \mathcal{F}_g\left[\exp\left(-\frac{\langle\underline{x},\underline{x}\rangle}{2}\right)\right](\underline{y}) = \mathcal{F}_g\left[\exp\left(-\frac{(x^j)^T G(x^j)}{2}\right)\right](y^j) \\
&= \frac{1}{\sqrt{\lambda_1\ldots\lambda_m}}\ \mathcal{F}\left[\exp\left(-\frac{(AP^{-1}(x'^j))^T\ G\ (AP^{-1}(x'^j))}{2}\right)\right](PA^T(y^j)) \\
&= \frac{1}{\sqrt{\lambda_1\ldots\lambda_m}}\ \mathcal{F}\left[\exp\left(-\frac{(x'^j)^T(x'^j)}{2}\right)\right](PA^T(y^j)) \\
&= \frac{1}{\sqrt{\lambda_1\ldots\lambda_m}}\ \exp\left(-\frac{(PA^T(y^j))^T\ PA^T(y^j)}{2}\right) \\
&= \frac{1}{\sqrt{\lambda_1\ldots\lambda_m}}\ \exp\left(-\frac{(y^j)^T G(y^j)}{2}\right) \\
&= \frac{1}{\sqrt{\lambda_1\ldots\lambda_m}}\ \exp\left(-\frac{\langle\underline{y},\underline{y}\rangle}{2}\right) = \frac{1}{\sqrt{\lambda_1\ldots\lambda_m}}\ \exp\left(-\frac{\|\underline{y}\|^2}{2}\right).
\end{aligned}$$

Hence, we indeed have

$$\mathcal{F}_g[\psi_t(\underline{x})](\underline{y}) = \frac{1}{\sqrt{\lambda_1\ldots\lambda_m}}\ (-i)^t\ \underline{y}^t\ \exp\left(-\frac{\|\underline{y}\|^2}{2}\right).$$

□

4.3. The g-Clifford-Hermite Continuous Wavelet Transform

In order to introduce the corresponding *g-Clifford-Hermite CWT*, we consider, still for $t > 0$, the continuous family of wavelets

$$\psi_t^{a,\underline{b}}(\underline{x}) = \frac{1}{a^{m/2}}\ \psi_t\left(\frac{\underline{x}-\underline{b}}{a}\right),$$

with $a \in \mathbb{R}_+$ the dilation parameter and $\underline{b} \in \mathbb{R}^m$ the translation parameter. The g-Clifford-Hermite CWT (g-CHCWT) applies to functions $f \in L_2(\mathbb{R}^m, dx^M)$ by

$$\begin{aligned}
T_t[f](a,\underline{b}) = F_t(a,\underline{b}) = \langle\psi_t^{a,\underline{b}}, f\rangle &= \int_{\mathbb{R}^m}\left(\psi_t^{a,\underline{b}}(\underline{x})\right)^\dagger\ f(\underline{x})\ dx^M \\
&= \frac{1}{a^{m/2}}\int_{\mathbb{R}^m}\exp\left(-\frac{\|\underline{x}-\underline{b}\|^2}{2a^2}\right)\ \left(H_t\left(\frac{\underline{x}-\underline{b}}{a}\right)\right)^\dagger\ f(\underline{x})\ dx^M.
\end{aligned}$$

This definition can be rewritten in the frequency domain as

$$\begin{aligned}
F_t(a,\underline{b}) &= \lambda_1\ldots\lambda_m\ \langle\mathcal{F}_g[\psi_t^{a,\underline{b}}],\mathcal{F}_g[f]\rangle \\
&= \lambda_1\ldots\lambda_m\int_{\mathbb{R}^m}\left(\mathcal{F}_g[\psi_t^{a,\underline{b}}](\underline{y})\right)^\dagger\ \mathcal{F}_g[f](\underline{y})\ dy^M.
\end{aligned}\tag{4.3}$$

The Fourier transform of the continuous family of wavelets:

$$\mathcal{F}_g[\psi_t^{a,\underline{b}}](\underline{y}) = \left(\frac{1}{\sqrt{2\pi}}\right)^m\int_{\mathbb{R}^m}\exp\left(-i\langle\underline{x},\underline{y}\rangle\right)\ \frac{1}{a^{m/2}}\ \psi_t\left(\frac{\underline{x}-\underline{b}}{a}\right)\ dx^M,$$

can be calculated by means of the successive substitutions $\underline{t} = \underline{x} - \underline{b}$ and $\underline{u} = \frac{\underline{t}}{a}$ yielding

$$\begin{aligned}
\mathcal{F}_g[\psi_t^{a,\underline{b}}](\underline{y}) &= \left(\frac{1}{\sqrt{2\pi}}\right)^m \frac{1}{a^{m/2}} \exp\left(-i\langle \underline{b}, \underline{y}\rangle\right) \int_{\mathbb{R}^m} \exp\left(-i\langle \underline{t}, \underline{y}\rangle\right) \psi_t\left(\frac{\underline{t}}{a}\right) dt^M \\
&= \left(\frac{1}{\sqrt{2\pi}}\right)^m a^{m/2} \exp\left(-i\langle \underline{b}, \underline{y}\rangle\right) \int_{\mathbb{R}^m} \exp\left(-i\langle \underline{u}, a\underline{y}\rangle\right) \psi_t(\underline{u})\, du^M \\
&= a^{m/2} \exp\left(-i\langle \underline{b}, \underline{y}\rangle\right) \mathcal{F}_g[\psi_t](a\underline{y}).
\end{aligned}$$

Consequently (4.3) becomes

$$\begin{aligned}
F_t(a, \underline{b}) &= \lambda_1 \ldots \lambda_m\, a^{m/2} \int_{\mathbb{R}^m} \exp\left(i\langle \underline{b}, \underline{y}\rangle\right) \left(\mathcal{F}_g[\psi_t](a\underline{y})\right)^\dagger \mathcal{F}_g[f](\underline{y})\, dy^M \\
&= \lambda_1 \ldots \lambda_m\, a^{m/2}\, (2\pi)^{m/2}\, \mathcal{F}_g\Big[\left(\mathcal{F}_g[\psi_t](a\underline{y})\right)^\dagger \mathcal{F}_g[f](\underline{y})\Big](-\underline{b}). \qquad (4.4)
\end{aligned}$$

It is clear that the g-CHCWT will map $L_2(\mathbb{R}^m, dx^M)$ into a weighted L_2-space on $\mathbb{R}_+ \times \mathbb{R}^m$ for some weight function still to be determined. This weight function has to be chosen in such a way that the CWT is an isometry, or in other words that the Parseval formula should hold.

Introducing the inner product

$$[F_t, Q_t] = \frac{1}{C_t} \int_{\mathbb{R}^m} \int_0^{+\infty} \left(F_t(a, \underline{b})\right)^\dagger Q_t(a, \underline{b}) \frac{da}{a^{m+1}}\, db^M,$$

we search for the constant C_t in order to have the Parseval formula

$$[F_t, Q_t] = \langle f, q\rangle$$

fulfilled.

By means of (4.4) and the Parseval formula for the g-Fourier transform (Theorem 2.49), we have consecutively

$$\begin{aligned}
[F_t, Q_t] &= \frac{(2\pi)^m}{C_t} (\lambda_1 \ldots \lambda_m)^2 \int_{\mathbb{R}^m} \int_0^{+\infty} \left(\mathcal{F}_g\left[\left(\mathcal{F}_g[\psi_t](a\underline{y})\right)^\dagger \mathcal{F}_g[f](\underline{y})\right](-\underline{b})\right)^\dagger \\
&\quad \mathcal{F}_g\left[\left(\mathcal{F}_g[\psi_t](a\underline{y})\right)^\dagger \mathcal{F}_g[q](\underline{y})\right](-\underline{b}) \frac{da}{a}\, db^M \\
&= \frac{(2\pi)^m}{C_t} (\lambda_1 \ldots \lambda_m)^2 \\
&\quad \int_0^{+\infty} \left\langle \mathcal{F}_g\left[\left(\mathcal{F}_g[\psi_t](a\underline{y})\right)^\dagger \mathcal{F}_g[f](\underline{y})\right], \mathcal{F}_g\left[\left(\mathcal{F}_g[\psi_t](a\underline{y})\right)^\dagger \mathcal{F}_g[q](\underline{y})\right]\right\rangle \frac{da}{a} \\
&= \frac{(2\pi)^m}{C_t} \lambda_1 \ldots \lambda_m \\
&\quad \int_0^{+\infty} \Big\langle \left(\mathcal{F}_g[\psi_t](a\underline{y})\right)^\dagger \mathcal{F}_g[f](\underline{y}), \left(\mathcal{F}_g[\psi_t](a\underline{y})\right)^\dagger \mathcal{F}_g[q](\underline{y})\Big\rangle \frac{da}{a}
\end{aligned}$$

$$= \frac{(2\pi)^m}{C_t} \lambda_1 \ldots \lambda_m$$

$$\int_{\mathbb{R}^m} \int_0^{+\infty} (\mathcal{F}_g[f](\underline{y}))^\dagger \ \mathcal{F}_g[\psi_t](a\underline{y}) \ (\mathcal{F}_g[\psi_t](a\underline{y}))^\dagger \ \mathcal{F}_g[q](\underline{y}) \ \frac{da}{a} \ dy^M$$

$$= \frac{(2\pi)^m}{C_t} \lambda_1 \ldots \lambda_m$$

$$\int_{\mathbb{R}^m} (\mathcal{F}_g[f](\underline{y}))^\dagger \left(\int_0^{+\infty} \mathcal{F}_g[\psi_t](a\underline{y}) \ (\mathcal{F}_g[\psi_t](a\underline{y}))^\dagger \ \frac{da}{a} \right) \mathcal{F}_g[q](\underline{y}) \ dy^M.$$

By means of the substitution

$$\underline{y} = \frac{r}{a} \ \underline{\omega}, \quad \underline{\omega} \in S^{m-1} \qquad \text{for which} \qquad \frac{da}{a} = \frac{dr}{r},$$

the integral between brackets becomes

$$\int_0^{+\infty} \mathcal{F}_g[\psi_t](a\underline{y}) \ (\mathcal{F}_g[\psi_t](a\underline{y}))^\dagger \ \frac{da}{a} = \int_0^{+\infty} \mathcal{F}_g[\psi_t](r\underline{\omega}) \ (\mathcal{F}_g[\psi_t](r\underline{\omega}))^\dagger \ \frac{dr}{r}.$$

As

$$\mathcal{F}_g[\psi_t](\underline{y}) \left(\mathcal{F}_g[\psi_t](\underline{y})\right)^\dagger = \frac{1}{\lambda_1 \ldots \lambda_m} \ \|\underline{y}\|^{2t} \ \exp\left(-\|\underline{y}\|^2\right)$$

is radial symmetric (i.e., only depending on $\|\underline{y}\|$), the integral between brackets can be further simplified to

$$\int_0^{+\infty} \mathcal{F}_g[\psi_t](a\underline{y}) \ (\mathcal{F}_g[\psi_t](a\underline{y}))^\dagger \ \frac{da}{a} = \frac{1}{A_m} \int_{\mathbb{R}^m} \frac{\mathcal{F}_g[\psi_t](\underline{y}) \left(\mathcal{F}_g[\psi_t](\underline{y})\right)^\dagger}{\|\underline{y}\|^m} \ dy^M$$

where A_m denotes the area of the unit sphere S^{m-1} in $\mathbb{R}^m$.

Consequently, if we put

$$\begin{aligned} C_t &= \frac{(2\pi)^m}{A_m} \int_{\mathbb{R}^m} \frac{\mathcal{F}_g[\psi_t](\underline{y}) \left(\mathcal{F}_g[\psi_t](\underline{y})\right)^\dagger}{\|\underline{y}\|^m} \ dy^M \\ &= \frac{(2\pi)^m}{\lambda_1 \ldots \lambda_m} \ \frac{(t-1)!}{2}, \end{aligned} \tag{4.5}$$

the desired Parseval formula follows:

$$\begin{aligned} [F_t, Q_t] &= \lambda_1 \ldots \lambda_m \int_{\mathbb{R}^m} \left(\mathcal{F}_g[f](\underline{y})\right)^\dagger \ \mathcal{F}_g[q](\underline{y}) \ dy^M \\ &= \lambda_1 \ldots \lambda_m \ \langle \mathcal{F}_g[f], \mathcal{F}_g[q] \rangle = \langle f, q \rangle. \end{aligned}$$

Hence we have proved that the g-Clifford-Hermite wavelets satisfy the so-called *admissibility condition*

$$C_t = \frac{(2\pi)^m}{A_m} \int_{\mathbb{R}^m} \frac{\mathcal{F}_g[\psi_t](\underline{y}) \ \left(\mathcal{F}_g[\psi_t](\underline{y})\right)^\dagger}{\|\underline{y}\|^m} \ dy^M < \infty.$$

The g-Clifford-Hermite CWT is thus an isometry between the spaces $L_2(\mathbb{R}^m, dx^M)$ and $L_2(\mathbb{R}_+ \times \mathbb{R}^m, C_t^{-1} a^{-(m+1)} \ da \ db^M)$.

References

[1] J.-P. Antoine, R. Murenzi, P. Vandergheynst and Syed Twareque Ali, *Two-Dimensional Wavelets and their Relatives*, Cambridge University Press, Cambridge, 2004.

[2] F. Brackx, R. Delanghe and F. Sommen, *Clifford Analysis*, Pitman Publishers (Boston-London-Melbourne, 1982).

[3] F. Brackx, R. Delanghe and F. Sommen, *Differential Forms and/or Multi-vector Functions*, CUBO A Mathematical Journal **7** (2005), no. 2, 139–169.

[4] F. Brackx, N. De Schepper and F. Sommen, *The Bi-axial Clifford Hermite Continuous Wavelet Transform*, Journal of Natural Geometry **24** (2003), 81–100.

[5] F. Brackx, N. De Schepper and F. Sommen, *The Clifford-Gegenbauer Polynomials and the Associated Continuous Wavelet Transform*, Integral Transform. Spec. Funct. **15** (2004), no. 5, 387–404.

[6] F. Brackx, N. De Schepper and F. Sommen, *The Clifford-Laguerre Continuous Wavelet Transform*, Bull. Belg. Math. Soc. - Simon Stevin **11**(2), 2004, 201–215.

[7] F. Brackx, N. De Schepper and F. Sommen, *Clifford-Jacobi Polynomials and the Associated Continuous Wavelet Transform in Euclidean Space* (accepted for publication in the Proceedings of the 4th International Conference on Wavelet Analysis and Its Applications, University of Macau, China, 2005).

[8] F. Brackx, N. De Schepper and F. Sommen, *New multivariable polynomials and their associated Continuous Wavelet Transform in the framework of Clifford Analysis* (submitted for publication in the Proceedings of the International Conference on Recent trends of Applied Mathematics based on partial differential equations and complex analysis, Hanoi, 2004).

[9] F. Brackx and F. Sommen, *Clifford-Hermite Wavelets in Euclidean Space*, Journal of Fourier Analysis and Applications **6**, no. 3 (2000), 299–310.

[10] F. Brackx and F. Sommen, *The Generalized Clifford-Hermite Continuous Wavelet Transform*, Advances in Applied Clifford Algebras **11(S1)**, 2001, 219–231.

[11] P. Calderbank, *Clifford analysis for Dirac operators on manifolds-with-boundary*, Max Planck-Institut für Mathematik (Bonn, 1996).

[12] C.K. Chui, *An Introduction to Wavelets*, Academic Press, Inc., San Diego, 1992.

[13] J. Cnops, *An introduction to Dirac operators on manifolds*, Birkhäuser Verlag (Basel, 2002).

[14] I. Daubechies, *Ten Lectures on Wavelets*, SIAM, Philadelphia, 1992.

[15] R. Delanghe, F. Sommen and V. Souček, *Clifford Algebra and Spinor-Valued Functions*, Kluwer Academic Publishers (Dordrecht, 1992).

[16] D. Eelbode and F. Sommen, *Differential Forms in Clifford Analysis* (accepted for publication in the Proceedings of the International Conference on Recent trends of Applied Mathematics based on partial differential equations and complex analysis, Hanoi, 2004).

[17] M. Felsberg, *Low-Level Image Processing with the Structure Multivector*, PhD-thesis, Christian-Albrechts-Universität, Kiel, 2002.

[18] J. Gilbert and M. Murray, *Clifford Algebras and Dirac Operators in Harmonic Analysis*, Cambridge University Press (Cambridge, 1991).

[19] K. Gürlebeck and W. Sprössig, *Quaternionic analysis and elliptic boundary value problems*, Birkhäuser Verlag (Basel, 1990).

[20] K. Gürlebeck and W. Sprössig, *Quaternionic and Clifford Calculus for Physicists and Engineers*, John Wiley & Sons (Chichester etc., 1997).

[21] G. Kaiser, *A Friendly Guide to Wavelets*, Birkhäuser Verlag (Boston, 1994).

[22] G. Kaiser, private communication.

[23] N. Marchuk, *The Dirac Type Tensor Equation in Riemannian Spaces*, In: F. Brackx, J.S.R. Chisholm and V. Souček (eds.), *Clifford Analysis and Its Applications*, Kluwer Academic Publishers (Dordrecht-Boston-London, 2001).

[24] M. Mitrea, *Clifford Wavelets, Singular Integrals and Hardy Spaces*, Lecture Notes in Mathematics 1575, Springer-Verlag (Berlin, 1994).

[25] T. Qian, Th. Hempfling, A. McIntosh and F. Sommen (eds.), *Advances in Analysis and Geometry: New Developments Using Clifford Algebras*, Birkhäuser verlag (Basel-Boston-Berlin, 2004).

[26] J. Ryan and D. Struppa (eds.), *Dirac operators in analysis*, Addison Wesley Longman Ltd, (Harlow, 1998).

[27] F. Sommen, *Special Functions in Clifford analysis and Axial Symmetry.* Journal of Math. Analysis and Applications **130** (1988), no. 1, 110–133.

Fred Brackx, Nele De Schepper and Frank Sommen
Clifford Research Group
Department of Mathematical Analysis
Faculty of Engineering – Faculty of Sciences
Ghent University
Galglaan 2
B-9000 Gent, Belgium

e-mail: fb@cage.ugent.be
e-mail: nds@cage.ugent.be
e-mail: fs@cage.ugent.be

Operator Theory:
Advances and Applications, Vol. 167, 69–85

A Hierarchical Semi-separable Moore-Penrose Equation Solver

Patrick Dewilde and Shivkumar Chandrasekaran

Abstract. The main result of the present paper is a method to transform a matrix or operator which has a hierarchical semi-separable (HSS) representation into a URV (Moore-Penrose) representation in which the operators U and V represent collections of efficient orthogonal transformations and the block upper matrix R still has the HSS form. The paper starts with an introduction to HSS-forms and a survey of a recently derived multi resolution representation for such systems. It then embarks on the derivation of the main ingredients needed for a Moore-Penrose reduction of the system while keeping the HSS structure. The final result is presented as a sequence of efficient algorithmic steps, the efficiency resulting from the HSS structure that is preserved throughout.

Mathematics Subject Classification (2000). Primary: 65F; Secondary: 15A.

Keywords. Hierarchically semi-separable systems, hierarchically quasi-separable systems, Moore-Penrose inverse, structured matrices.

1. Introduction

Many physical systems are modeled by systems of differential equations, integral equations or combinations of them. Solving these systems requires discretization of the equations and leads to large systems of algebraic equations. In the case of systems governed by linear equations, the resulting system of equations will be linear as well and can be solved either directly or iteratively, involving a preconditioner and a Lanczos-type recursion. In the case of non-linear systems, the discretized system will be non-linear as well and an iterative procedure has to be set up to find the solution. Such a procedure is, e.g., of the 'Newton-Raphson' type and would in turn require the solution of a system of linear equations, now involving a differential such as a Jacobian. In all cases the resulting systems tend to lead to matrices of very large dimensions, even for fairly small problems, so that solvers using standard numerical procedures quickly run out of steam. Luckily, many systems exhibit quite a bit of structure that can be exploited to make the solver more efficient. In the case of a discretized system for a differential equation, the resulting matrix structure will be very sparse, as only entries corresponding to nearby points will be different from zero. In this case, the matrix-vector multiplication can be efficiently executed. Iterative methods are well suited to exploit this fact, but they are dependent on a low-complexity approximant of the inverse of the original system, the so-called preconditioner. Approximate solutions of the system can be iteratively constructed via low-complexity calculations, provided the pre-conditioner exhibits the necessary structure.

In the case of integral equations, it has been remarked by Gohberg, Kailath and Koltracht [9] and Rokhlin [12] that low rank approximations of the integral kernel lead to large submatrices of low rank in the resulting system of equations. Exploiting this structure which was termed 'Semi-Separable' leads to solution procedures that are linear in the size of the matrix and quadratic in the size of the approximation. A systematic method to obtain such low rank approximations was proposed by Greengard and Rokhlin [10] and is known as the Fast Multipole Method. These original approaches suffered from numerical problems as the use of backward stable orthogonal transformations in this context was not yet well understood. The introduction of time-varying system theory to model the system of equations [13] provided for the necessary structure to allow for more general types of transformations than those used by the original authors cited. A survey of these techniques can be found in the book [6]. Based on these ideas, the Semi-Separable structure was extended to a more generic form called 'Quasi-Separable' and numerically stable system solvers were developed for this structure by a number of authors [8, 3, 7].

Although these developments lead to a satisfactory and useful theory, it was also evident that they did not exploit the structural properties of most physical systems sufficiently. Two examples may suffice to illustrate this point. In the case of a partial differential equation in 3D space, discretization coordinates will have three indices, say $\{i, j, k\}$, and interaction between values in close-by points may be expected. To construct the interaction matrix, each discretization coordinate has to be assigned a single index. In case of a regular grid of dimension N^3, the index assignment would run as $i + Nj + N^2k$ and the resulting matrix would have a hierarchical structure consisting of a diagonal bands of blocks of dimension N^2 each consisting of diagonal bands of blocks of dimension N, which in turn consist of scalar diagonal bands. The Semi-Separable or Quasi-Separable theory is insufficient to handle such types of matrices, it gets a good grip only on the top level of the hierarchy, while the structure of the lower hierarchical levels is greatly disturbed [5]. Also in the case of the multipole method and assuming the distribution of 'objects' handled by the method in 3D space to be fairly general (assuming of course that the multipole assumption holds as well), a similar problem will arise: many submatrices will be of low rank, but they will have a 'hierarchical ordering', restricting the applicability of the Semi-Separable method.

In the present paper we deal with an intermediate structure which has a nice hierarchical (or equivalently multi-resolution) structure and is capable to cope with the problem mentioned in the previous paragraph. The structure was first presented in [4], and a few solvers for it were presented [11]. In particular, [2] shows how the structure can be reduced to a sparse, directly solvable system, using a state space model of intermediate variables, much as was done for the Semi- or Quasi-separable case. In contrast to the latter the state-space model turns out to have a hierarchical (multi-resolution) structure, which can efficiently be exploited. These straight solvers assume the original system to be square non-singular, allowing for partial recursive elimination of unknowns and recursive back-substitutions as the algorithm proceeds. In the present paper we propose a new, backward stable solver that finds the Moore-Penrose solution for a general system of equations, namely a system that is not assumed to be square, non-singular. Our goal is to obtain the same order of numerical complexity as the straight solvers, but now for the Moore-Penrose case.

2. HSS representations

The Hierarchical Semi-Separable representation of a matrix A is a layered representation of the multi-resolution type, indexed by the hierarchical level. At the top level 1, it is a 2×2 block matrix representation of the form:

$$A = \begin{bmatrix} A_{1;1,1} & A_{1;1,2} \\ A_{2;2,1} & A_{2;2,2} \end{bmatrix} \tag{2.1}$$

in which we implicitly assume that the ranks of the off-diagonal blocks is low so that they can be represented by an 'economical' factorization ('H' indicates Hermitian transposition, for real matrices just transposition), as follows:

$$A = \begin{bmatrix} D_{1;1} & U_{1;1}B_{1;1,2}V_{1;2}^H \\ U_{1;2}B_{1;2,1}V_{1;1}^H & D_{1;2} \end{bmatrix}. \tag{2.2}$$

The second hierarchical level is based on a further but similar decomposition of the diagonal blocks, respect. $D_{1;1}$ and $D_{1;2}$:

$$\begin{aligned} D_{1;1} &= \begin{bmatrix} D_{2:1} & U_{2;1}B_{2;1,2}V_{2;2}^H \\ U_{2;2}B_{2;2,1}V_{2;1}^H & D_{2;2} \end{bmatrix}, \\ D_{1;2} &= \begin{bmatrix} D_{2;3} & U_{2;3}B_{2;3,4}V_{2;4}^H \\ U_{2;4}B_{2;4,3}V_{2;3}^H & D_{2;4} \end{bmatrix} \end{aligned} \tag{2.3}$$

for which we have the further *level compatibility* assumption

$$\operatorname{span}(U_{1;1}) \subset \operatorname{span}\left(\begin{bmatrix} U_{2;1} \\ 0 \end{bmatrix}\right) \oplus \operatorname{span}\left(\begin{bmatrix} 0 \\ U_{2;2} \end{bmatrix}\right), \tag{2.4}$$

$$\operatorname{span}(V_{1;1}) \subset \operatorname{span}\left(\begin{bmatrix} V_{2;1} \\ 0 \end{bmatrix}\right) \oplus \operatorname{span}\left(\begin{bmatrix} 0 \\ V_{2;2} \end{bmatrix}\right) \text{ etc} \ldots \tag{2.5}$$

This spanning property is characteristic for the HSS structure, it is a kind of hierarchical 'Lanczos' property and allows a substantial improvement on the numerical complexity as a multiplication with higher level structures always can be done using lower level multiplications, using so called 'translation operators'

$$U_{1;i} = \begin{bmatrix} U_{2;2i-1}R_{2;2i-1} \\ U_{2;2i}R_{2;2i} \end{bmatrix}, \quad i = 1, 2, \tag{2.6}$$

$$V_{1;i} = \begin{bmatrix} V_{2;2i-1}W_{2;2i-1} \\ V_{2;2i}W_{2;2i} \end{bmatrix}, \quad i = 1, 2. \tag{2.7}$$

Notice the use of indices: at a given level i rows respect. columns are subdivided in blocks indexed by $1, \ldots, i$. Hence the ordered index $(i; k, \ell)$ indicates a block at level i in the position (k, ℓ) in the original matrix. The same kind of subdivision can be used for column vectors, row vectors and bases thereof (as are generally represented in the matrices U and V).

In [2] it is shown how this multilevel structure leads to efficient matrix-vector multiplication and a set of equations that can be solved efficiently as well. For the sake of completeness we review this result briefly here. Let us assume that we want to solve the system $Ax = b$ and that A has an HSS representation with deepest hierarchical level K. We begin by accounting for the matrix-vector multiplication Ax. At the leave node $(K; i)$ we can compute

$$g_{K;i} = V_{K;i}^H x_{K;i}.$$

If $(k; i)$ is not a leaf node, we can infer, using the hierarchical relations

$$g_{k;i} = V_{k;i}^H x_{k;i} = W_{k+1;2i-1}^H g_{k+1;2i-1} + W_{k+1;2i}^H g_{k+1;2i}.$$

These operations update a 'hierarchical state' $g_{k;i}$ upwards in the tree. To compute the result of the multiplication, a new collection of state variables $\{f_{k;i}$ is introduced for which it holds that

$$b_{k;i} = A_{k;i,i} + U_{k;i} f_{k;i}$$

and which can also be computed recursively downwards by the equations

$$\begin{bmatrix} f_{k+1;2i-1} \\ f_{k+1;2i} \end{bmatrix} = \begin{bmatrix} B_{k+1;2i-1,2i} g_{k+1;2i} + R_{k+1;2i-1} f_{k,i} \\ B_{k+1;2i,2i-1} g_{k+1;2i-1} + R_{k+1;2i} f_{k;i} \end{bmatrix},$$

the starting point being $f_{0;} = []$, an empty matrix. At the leaf level we can now compute (at least in principle - as we do not know x) the outputs from

$$b_{K;i} = D_{K;i} x_{K;i} + U_{K;i} f_{K;i}.$$

The next step is to represent the multiplication recursions in a compact form using matrix notation and without indices. We fix the maximum order K as before. Next we define diagonal matrices containing the numerical information, in breadth first order:

$$\mathbf{D} = \mathrm{diag}[D_{K;i}]_{i=1,\ldots,K}, \ \mathbf{W} = \mathrm{diag}[(W_{1;i})_{i=1,2}, (W_{2;i})_{i=1\cdots 4}, \ldots], \ \text{etc}\ldots$$

Next, we need two shift operators relevant for the present situation, much as the shift operator Z in time-varying system theory [6]. The first one is the shift-down operator $Z_{\downarrow}$ on a tree. It maps a node in the tree on its children and is a nilpotent operator. The other one is the level exchange operator $Z_{\leftrightarrow}$. At each level it exchanges children of the same node and is a permutation operator. Finally, we need the leaf projection operator $\mathbf{P}_{\text{leaf}}$ which on a state vector which assembles in breadth first order all the values $f_{k;i}$ produces the values of the leaf nodes (again in breadth first order). The state equations representing the efficient multiplication can now be written as

$$\left\{\begin{array}{lllll} \mathbf{g} & = & \mathbf{P}_{\text{leaf}}^H \mathbf{V}^H \mathbf{x} & + & Z_{\downarrow}^H \mathbf{W}^H \mathbf{g} \\ \mathbf{f} & = & \mathbf{R} Z_{\downarrow} \mathbf{f} & + & \mathbf{B} Z_{\leftrightarrow} \mathbf{g} \end{array}\right. \tag{2.8}$$

while the 'output' equation is given by

$$\mathbf{b} = \mathbf{D}\mathbf{x} + \mathbf{U}\mathbf{P}_{\text{leaf}}\mathbf{f}. \tag{2.9}$$

This system of equations is sparse and can always be solved (even efficiently, that is by visiting the given data once), because $(I - \mathbf{W}Z_{\downarrow})$ and $(I - \mathbf{R}Z_{\downarrow})$ are invertible operators due to the fact that $Z_{\downarrow}$ is nilpotent. We obtain

$$A = \mathbf{D} + \mathbf{U}\mathbf{P}_{\text{leaf}}(I - \mathbf{R}Z_{\downarrow})^{-1}\mathbf{B}(I - Z_{\downarrow}^H \mathbf{W}^H)^{-1}\mathbf{P}_{\text{leaf}}\mathbf{V}^H)\mathbf{x} = \mathbf{b}. \tag{2.10}$$

Various strategies can be used to solve this sparse system of equations, we refer to the paper mentioned for more information. One elimination procedure that is aesthetically attractive follows the hierarchical ordering of the data bottom up. In a tree that is two levels down the elimination order would be:

$$\begin{array}{c} (f_{2;1}, g_{2;1}, x_{2;1}), (f_{2;2}, g_{2;2}, x_{2;2}), (f_{2;3}, g_{2;3}, x_{2;3}), (f_{2;4}, g_{2;4}, x_{2;4}), \\ (f_{1;1}, g_{1;1}), (f_{1;2}, g_{1;2}), (f_{0;1}, g_{0;1}). \end{array} \tag{2.11}$$

The computation must start at the leaf nodes, where multiplication with the base vectors takes place, in higher up locations there is only multiplication with transfer operators which relate the higher up bases to the bases at the leaf level. In the paper cited it is shown that this procedure hierarchically eliminates unknowns without producing any fill ins in the original sparse matrix describing the system.

The present paper aims at presenting a QR-type elimination procedure capable of deriving the Moore-Penrose inverse of a (presumably singular or ill conditioned) problem. The additional difficulty here is that elimination of variables cannot be done on the fly, because the Moore-Penrose solution can only be determined after the whole structure has been visited. Therefore we will aim at constructing the Moore-Penrose inverse rather than at solving the equations recursively as they appear.

3. Preliminaries

We shall use a number of reduction theorems (in a well-specified order).

Proposition 1. *Let* $V = \left[\begin{array}{c} V_1 \\ \hline V_2 \end{array}\right]$ *be a (tall) full rank matrix of size* $(k+m) \times k$ *with* $m \geq k$, *then an orthogonal transformation* Q *that reduces* V *to the form*

$$QV = \left[\begin{array}{c} 0 \\ \hline R \end{array}\right] \tag{3.12}$$

with R square non-singular exists and can be chosen of the form

$$Q = \left[\begin{array}{c|c} w_1 & -L^H w_2^H \\ \hline w_2 & I - w_2 K w_2^H \end{array}\right] \tag{3.13}$$

in which $w_1 = V_1 R^{-1}$, $w_2 = V_2 R^{-1}$, K is a hermitian matrix that satisfies

$$I - w_2 K w_2^H = (I - w_2 w_2^H)^{1/2} \tag{3.14}$$

and L a unitary matrix that satisfies

$$(I - K w_2^H w_2) = w_1 L. \tag{3.15}$$

Proof. The theorem claims the existence of K and L implicitly. If V_1 happens to be invertible, then this is straightforward, the difficulty is when V_1 is not invertible, we use an implicit proof to cover the general case. R is defined as a square matrix that satisfies

$$V_1^H V_1 + V_2^H V_2 = R^H R \tag{3.16}$$

and its non-singularity follows from the non-singularity assumption on V. Next let $w_1 = V_1 R^{-1}$ and $w_2 = V_2 R^{-1}$, then $w = \left[\begin{array}{c} w_1 \\ \hline w_2 \end{array}\right]$ is isometric, $w_1^H w_1$ is contractive and an eigenvalue decomposition

$$w_1^H w_1 = v_1 \sigma^2 v_1^H \tag{3.17}$$

can be chosen such that the positive diagonal matrix σ satisfies $0 \leq \sigma \leq I$. Since w_1 is square, a unitary u_1 will exist such that $w_1 = u_1 \sigma v_1^H$ (the proof goes as in the proof of the SVD). Next, $w_2^H w_2 = I - w_1^H w_1$ and an eigenvalue decomposition for it is

$$v_1 (I - \sigma^2) v_1^H. \tag{3.18}$$

Since w_2 is tall by assumption, there will exist u_2 isometric such that

$$w_2 = u_2 (I - \sigma^2)^{1/2} v_1^H. \tag{3.19}$$

It is now easy to verify directly that

$$Q = \left[\begin{array}{c|c} u_1 \sigma v_1^H & -u_1 (I - \sigma^2)^{1/2} u_2^H \\ \hline u_2 (I - \sigma^2)^{1/2} v_1^H & I - u_2 (I - \sigma) u_2^H \end{array}\right] \tag{3.20}$$

is a unitary matrix. Putting $K = v(I+\sigma)^{-1} v^H$ and $L = v u_1^H$ produces the desired form of Q. The converse check that any Q with the given form is unitary is also immediate by direct verification. □

The theorem shows that a unitary matrix close to identity (where 'close' means 'the difference is a low rank matrix') can be constructed that reduces a tall matrix to a small triangular matrix. In Numerical Analysis one traditionally uses 'Householder transformations' for this purpose, the transformation presented here has the advantage that its determinant can be controlled more easily.

Proposition 2. *Let* $T = \left[\begin{array}{c} UV^H \\ \hline D \end{array}\right]$ *in which* U *is a tall, isometric matrix of rank* δ, T *is of dimension* $(k+m) \times m$, *accordingly partitioned and of full column rank, and* $k \leq m$. *Let* $N^H N = VV^H + D^H D$, *in which* N *is a square matrix. Then* N *is non-singular, and there exists a unitary matrix* Q *such that*

$$T = Q \left[\begin{array}{c} 0 \\ \hline N \end{array}\right]. \tag{3.21}$$

Moreover, Q *can be chosen of the form*

$$Q = \left[\begin{array}{c|c} d_1 & Uv_r^H \\ \hline u_\ell v_\ell^H & d_2 \end{array}\right] \tag{3.22}$$

in which u_ℓ *has at most the same rank as* U *and* d_1 *is a rank* δ *perturbation of the unit matrix.*

Proof. The non-singularity of N follows directly from the full column rank assumption. The proof is then based on 'joint' SVD's of $UV^H N^{-1}$ and DN^{-1} and then completing the form. More precisely, let $N^{-T}VUU^H V^H N^{-1} = v\sigma^2 v^H$ be an eigenvalue decomposition with unitary v and a positive diagonal matrix σ of dimension $m \times m$. Let $\pi_k = \left[\begin{array}{c|c} I_k & 0_{m-k} \end{array}\right]$. Then there will exist unitary matrices u_1 and u_2 so that

$$\left[\begin{array}{c} UV^H N^{-1} \\ \hline DN^{-1} \end{array}\right] = \left[\begin{array}{c} u_1 \pi_k \sigma v^H \\ \hline u_2 (I - \sigma^2)^{1/2} v^H \end{array}\right]. \tag{3.23}$$

(It is not hard to check that the right-hand side form is indeed isometric!) Let moreover $\sigma_k = \pi_k \sigma \pi_k^H$. A Q that satisfies the requirements of the theorem is now easily constructed as

$$Q = \left[\begin{array}{c|c} u_1 (I - \sigma_k^2)^{1/2} u_1^H & u_1 \pi_k \sigma v^H \\ \hline -u_2 \sigma \pi_k^H u_1^H & u_2 (I - \sigma^2)^{1/2} v^H \end{array}\right]. \tag{3.24}$$

As the rank of σ is at most equal to the rank of U and $UV^H N^{-1} = u_1 \pi_k \sigma v^H$ there exist v_r such that $u_1 \pi_k \sigma v^H = Uv_r^H$. Finally, $(I - \sigma_k^2)^{1/2} = I - \sigma_\delta$ for some positive diagonal matrix σ_δ of rank at most δ – as σ_k itself has rank at most δ. Hence d_1 is at most a matrix of rank δ different from the unit matrix of dimension k. □

The exact form of the reduction matrices (the 'Q'-matrices in the procedure) is of importance for the complexity of the algorithms that will be derived next, because they have to be propagated to other submatrices than the one they are originally derived from. We shall measure computation complexity by tallying how many times data items are visited. If an $m \times n$ matrix multiplies a vector of dimension n with a direct algorithm, then the complexity by this measure is exactly $m * n$, all data items in the matrix are visited once (the operation tally is higher because each multiplication is followed by an addition except the last). The complexity of the row absorption procedure of the previous theorem is δm^2

where δ is the number of columns in the U or V matrix, because the result can be obtained through a QR factorization on the matrix $\left[\begin{array}{c} V^H \\ \hline D \end{array}\right]$.

4. HSS row absorption procedure

Our next goal is to derive a so-called 'HSS row absorption' matrix – one of the main ingredients in the HSS Moore-Penrose reduction process. Starting point is the form

$$\left[\begin{array}{c|c} w_1 v_\ell^H & w_2 v_r^H \\ \hline\hline d_1 & u_r v_r^H \\ \hline u_\ell v_\ell^H & d_2 \end{array}\right] \tag{4.25}$$

in which we assume the top entries to be 'skinny', i.e., of low rank compared to the dimensions of the matrix (this assumption is not used explicitly but underlies the usefulness of the procedure – as discussed above, the HSS compatibility has to be preserved!). We introduce a somewhat simplified notation, whenever clear from the context we drop the level indication. E.g., in the above matrix, the notation U_r and $U_{1;r}$ or $U_{1;1}$ would be equivalent. The goal of the procedure is to reduce the matrix to a row independent HSS form using orthogonal transformations. Important in the reduction procedure are the properties of the overall orthogonal reducing matrix, namely which block entries in it have an HSS form and which are 'skinny' – i.e., have low rank of the same order as the rank of the top entries to be absorbed. The procedure uses the properties derived in the first section.

Step 1. Find orthogonal q so that

$$\left[\begin{array}{c|c} q_{11} & q_{12} \\ \hline q_{21} & q_{22} \end{array}\right] \left[\begin{array}{c} w_2 \\ \hline u_r \end{array}\right] = \left[\begin{array}{c} r \\ \hline 0 \end{array}\right] \tag{4.26}$$

with r square non-singular and q_{22} close to a unit matrix, and apply the transformation to the original after embedding:

$$\left[\begin{array}{c|c} q & \\ \hline & I \end{array}\right] \left[\begin{array}{c|c} w_1 v_\ell^H & w_2 v_r^H \\ \hline d_1 & u_r v_r^H \\ \hline\hline u_\ell v_\ell^H & d_2 \end{array}\right] = \left[\begin{array}{c|c} q_{11} w_1 v_\ell^H + q_{12} d_1 & r v_r^H \\ \hline q_{21} w_1 v_\ell^H + q_{22} d_1 & 0 \\ \hline\hline u_\ell v_\ell^H & d_2 \end{array}\right]. \tag{4.27}$$

Let

$$d_1' = q_{21} w_1 v_\ell^H + q_{22} d_1 \tag{4.28}$$

the product of a lower level form (d_1) with a 'skinny' perturbation of the unit matrix (q_{22}) followed by a 'skinny' additive perturbation ($q_{21} w_1 v_\ell^H$). The new look of the matrix is, after an exchange of block-rows

$$\left[\begin{array}{c|c} d_1' & 0 \\ \hline\hline q_{11} w_1 v_\ell^H + q_{12} d_1 & r v_r^H \\ \hline u_\ell v_\ell^H & d_2 \end{array}\right] \tag{4.29}$$

in which the product $q_{12}d_1$ is 'skinny', but increases the rank of that term beyond the rank of v_ℓ – as can be expected.

Step 2. We now work on the right part of the matrix. Let p be an orthogonal matrix that reduces

$$\left[\frac{rv_r^H}{d_2}\right]. \tag{4.30}$$

Since d_2 is square (this assumption is not really necessary!) the result will have the form

$$\left[\frac{0}{d_2'}\right]. \tag{4.31}$$

This procedure amounts to a lower level HSS row absorption problem (p cannot be claimed to be 'skinnily' away from a unit, it will have whatever structure it inherits from the lower level operation, which is isomorphic to what is happening at this level). Applying p to the left column will produce

$$p\left[\frac{q_{11}w_1v_\ell^H + q_{12}d_1}{u_\ell v_\ell^H}\right] = \left[\frac{v'^T}{u_\ell' v_\ell'^T}\right]. \tag{4.32}$$

The matrix has now been brought to the form

$$\left[\begin{array}{c|c} d_1' & 0 \\ \hline\hline v'^T & 0 \\ \hline u_\ell' v_\ell'^T & d_2' \end{array}\right]. \tag{4.33}$$

Now, if the original matrix is non-singular (has full column rank), then d_2' will have full row rank (also by construction) and a further, lower level absorption is needed (using a new transformation matrix s) on

$$\left[\frac{d_1'}{v'^T}\right] \tag{4.34}$$

to yield the end result in the form

$$\left[\begin{array}{c|c} d_1'' & 0 \\ \hline 0 & 0 \\ \hline u_\ell' v_\ell'^T & d_2' \end{array}\right], \tag{4.35}$$

in which both d_1'' and d_2' have full row rank (and in case the original matrix was non-singular, also full column rank and hence will be square). It can be remarked that however one looks at it, the full matrix has to be involved in the procedure, but all operations are either elementary or absorptions at a lower level or involve a very skinny transformation at the higher level (the q matrix).

Collecting the transformations we find for the overall Q (the fat entries are full, lower level matrices, q_{22} is skinnily away from unitary and the non-fat entries are skinny):

$$Q = \left[\begin{array}{c|c|c} \mathbf{s}_{11} & s_{12} & \\ \hline s_{21} & s_{22} & \\ \hline & & I \end{array}\right] \left[\begin{array}{c|c|c} I & & \\ \hline & p_{11} & p_{12} \\ \hline & p_{21} & \mathbf{p}_{22} \end{array}\right] \left[\begin{array}{c|c|c} q_{21} & \mathbf{q}_{22} & \\ \hline q_{11} & q_{12} & \\ \hline & & I \end{array}\right]. \tag{4.36}$$

Working out (should not be done in practice) produces:

$$Q = \left[\begin{array}{c|c|c} \mathbf{s}_{11}q_{21} + s_{12}p_{11}q_{11} & \mathbf{s}_{11}\mathbf{q}_{22} + s_{12}p_{11}q_{12} & s_{12}p_{12} \\ \hline s_{21}q_{21} + s_{22}p_{11}q_{12} & s_{21}\mathbf{q}_{22} + s_{22}p_{11}q_{22} & s_{22}p_{12} \\ \hline p_{21}q_{11} & p_{21}q_{12} & \mathbf{p}_{22} \end{array}\right]. \tag{4.37}$$

The most critical term is the 1,2 where the product of two full matrices occurs: $\mathbf{s}_{11}\mathbf{q}_{22}$. But $\mathbf{q}_{22}$ is only 'skinnily' away from a unit matrix, hence this product also has a 'skinny' algorithm. The final form for the matrix Q produces:

$$\left[\begin{array}{c|c|c} Q_{11} & \mathbf{Q}_{12} & Q_{13} \\ \hline Q_{21} & Q_{22} & Q_{23} \\ \hline Q_{31} & Q_{32} & \mathbf{Q}_{33} \end{array}\right] \left[\begin{array}{c|c} w_1 v_\ell^H & w_2 v_r^H \\ \hline d_1 & u_r v_r^H \\ u_\ell v_\ell^H & d_2 \end{array}\right] = \left[\begin{array}{c|c} d_1'' & 0 \\ \hline 0 & 0 \\ \hline u_\ell' v_\ell'^T & d_2' \end{array}\right] \tag{4.38}$$

in which the not-boldface entries of Q have low rank (skinny products), the boldface ones are of HSS form, and both d_1'' and d_2' have full column rank, and d_1'', d_2' have full row rank.

Complexity calculation

As the absorption procedure turns out to be the main workhorse in the reduction procedure, we proceed to its complexity tally. Let us assume that the vectors to be absorbed are of dimension δ_1, while the off-blocks in the lower HSS representation are of (effective) rank $\delta_2 \geq \delta_1$. Let $\mathbf{M}(n, \delta)$ indicate the computational complexity of multiplying an HSS matrix of dimension n with a block of δ vectors, and $\mathbf{C}(n, \delta)$ the complexity of the absorption procedure of a vector block of dimension δ by a dimension n HSS matrix. The complexity of the HSS absorption procedure can now be tallied. An important observation is that the computation of $q_{11}w_1v_\ell^H$ in this step can be postponed until the bottom level of the procedure is obtained. Let us assume that this bottom level is characterized by matrices of dimensions $N_{\text{bot}} \times N_{\text{bot}}$, then the complexity count would be as follows:

Step 1: $\delta_1 N_{\text{bot}} + 2\mathbf{M}(\frac{n}{2}, \delta_1)$;
Step 2: $\mathbf{C}(\frac{n}{2}, \delta_1) + \mathbf{M}(\frac{n}{2}, 2\delta_1) = \mathbf{C}(\frac{n}{2}, \delta_1) + 2\mathbf{M}(\frac{n}{2}, \delta_1)$;
Step 3: $\mathbf{C}(\frac{n}{2}, 2\delta_1) = 2\mathbf{C}(\frac{n}{2}, \delta_1)$.

Hence the total tally is

$$\delta_1 N_{\text{bot}} + 4\mathbf{M}(\frac{n}{2}, \delta_1) + 3\mathbf{C}(\frac{n}{2}, \delta_1). \tag{4.39}$$

We see that the complexity is not directly dependent on the top level dimension n, and just linearly on the lower level dimensions, where presumably similar computations will take place, to be shown in the next section. It is of course dependent on the local rank (δ_1) but also in a linear fashion. Another important point is in what state the matrix is left behind after the reduction procedure, more precisely whether the HSS relations with the other submatrices in the original are still valid. This point will be taken up in the next section where we consider the overall procedure.

The dual of the absorption will be needed as well in the sequel. At this point there is no substantial difference between the two procedures, one works on the rows, the other on the columns of the HSS matrix, producing a low complexity calculation that preserves the HSS structure.

5. An HSS Moore-Penrose reduction method

For ease of discussion, we assume that the system is a 'flat' system of full row rank, and furthermore given in the traditional HSS form. In case the assumption does not hold then the later steps in the algorithm will have to be modified, but this would not entail major difficulties. We start out with a matrix in HSS form, i.e., it is of the form

$$\left[\begin{array}{c|c} D_1 & U_u V_u^H \\ \hline U_\ell V_\ell^H & D_2 \end{array}\right] \tag{5.40}$$

in which the low rank matrices U_u, V_u, U_ℓ, V_ℓ are *HSS compatible*, i.e., can be generated from the lower HSS levels, as explained in the introduction. To keep low complexity calculations it will be imperative to preserve the HSS compatibility structure whenever appropriate. The first step is the replacement of the original HSS problem (MPHSS) by an equivalent modified set of equations as follows:

$$\left[\begin{array}{c|c} D_1 & U_u V_u^H \\ \hline U_\ell V_\ell^H & D_2 \end{array}\right] \Rightarrow \left[\begin{array}{c|cc} D_1 - D_1 V_\ell V_\ell^H & D_1 V_\ell & U_u V_u^H \\ \hline 0 & U_\ell & D_2 \end{array}\right]. \tag{5.41}$$

Here we assume that V_ℓ is a 'skinny' orthonormal set of columns and that D_1 and D_2 possibly again have HSS forms of lower hierarchical level. Before discussing the merits of this step, and then the further steps to be executed, we verify the algebraic correctness of this first step. Let V'_ℓ be an orthogonal complementary set of columns of V_ℓ, then postmultiplying the original system with an appropriately dimensioned orthogonal matrix embedding $[V'_\ell \; V_\ell]$ produces the equivalent system (the second member has to be adapted in an obvious way, we skip this step):

$$\left[\begin{array}{c|c} D_1 & U_u V_u^H \\ \hline U_\ell V_\ell^H & D_2 \end{array}\right] \left[\begin{array}{c|c} V'_\ell & V_\ell \\ \hline & I \end{array}\right] = \left[\begin{array}{c|cc} D_1 V'_\ell & D_1 V_\ell & U_u V_u^H \\ \hline 0 & U_\ell & D_2 \end{array}\right] \tag{5.42}$$

$$(D_1 - D_1 V_\ell V_\ell^H) V'_\ell = D_1 V'_\ell. \tag{5.43}$$

The 1,1 block entry $D_1 V'_\ell$ may now be replaced by $D_1(I - V_\ell V_\ell^H)$ (increasing the size of the matrix) because the MP solutions of the two systems are closely related: if $\begin{bmatrix} y_1 \\ y_2 \end{bmatrix}$ solves the latter, then $\begin{bmatrix} V'_\ell y_1 \\ y_2 \end{bmatrix}$ will solve the former, due to the fact that $V_\ell'^T V'_\ell = I$. The increase in size will be removed in later 'absorption' steps. In this step, D_1 gets modified by a matrix that is only 'skinnily' away from the unit matrix.

Before proceeding, we study the effects $(I - V_\ell V_\ell^H)$ has on D_1 and whether HSS compatibility is preserved in this step. Hence, we assume that D_1 has in turn a lower level HSS form. To avoid a surfeit of indices, we replace V_ℓ by V and express the lower level HSS form again with the same notation as before:

$$D_1 \leftarrow \left[\begin{array}{c|c} D_1 & U_r V_r^H \\ \hline U_\ell V_\ell^H & D_2 \end{array}\right]. \tag{5.44}$$

Furthermore, because of the HSS relations, we have

$$V = \left[\begin{array}{c} V_1 \\ \hline V_2 \end{array}\right] = \left[\begin{array}{c} V_\ell W_\ell \\ \hline V_r W_r \end{array}\right] \tag{5.45}$$

(which could also be expressed as

$$V = \left[\begin{array}{c} V_\ell \\ \hline V_r \end{array}\right] \odot \left[\begin{array}{c} W_\ell \\ \hline W_r \end{array}\right] \tag{5.46}$$

in which the W's are assumed tall matrices, and the $\odot$ indicates pointwise multiplication of block entries – assuming matching dimensions of course). We find as result for this operation

$$\left[\begin{array}{c|c} D_1 - (D_1 V_1 + U_r V_r^H V_2) V_1^H & U_r V_r^H - (D_1 V_1 + U_r V_r^H V_2) V_2^H \\ \hline U_\ell V_\ell^H - (U_\ell V_\ell^H V_1 + D_2 V_2) V_1^H & D_2 - (U_\ell V_\ell^H V_1 + D_2 V_2) V_2^H \end{array}\right]. \tag{5.47}$$

We see that the ranks of the off-diagonal terms have not increased, but the column basis has changed, and it may be useful to keep the original column vectors although they may not amount to a basis (they are able to generate the column vectors though). Taking this conservative approach we may write the result so far as

$$\left[\begin{array}{c|c} D_1' & [U_{rn}\ U_r] W_{rn}^H V_r^H \\ \hline [U_{\ell n}\ U_\ell] W_{\ell n}^H V_\ell^H & D_2' \end{array}\right] \tag{5.48}$$

in which

$$\left\{\begin{array}{rcl} U_{rn} & = & D_1 V_1 \\ D_1' & = & D_1 - (U_{rn} + U_r V_r^H V_2) V_1^H \\ W_{rn}^H & = & \left[\begin{array}{c} -W_r^H \\ I - V_r^H V_2 W_r^H \end{array}\right] \\ D_2' & = & D_2 - (U_{\ell n} + U_\ell V_\ell^H V_1) V_2^H \\ U_{\ell n} & = & D_2 V_2 \\ W_{\ell n}^H & = & \left[\begin{array}{c} -W_\ell^H \\ I - V_\ell^H V_1 W_\ell^H \end{array}\right] \end{array}\right. \tag{5.49}$$

all quantities that can now be computed at the lowest level, and are either 'skinny' or 'skinny' updates.

Going back to the original procedure, the same strategy can now be applied to the rightmost block in eq. 5.42, this produces:

$$\begin{array}{l} \left[\begin{array}{c|c} U_{1;r} V_{1;r,3}^H & U_{1,r} V_{1;r,4}^H \\ \hline\hline D_{2;3} & U_{2;r} V_{2;r}^H \\ \hline U_{2;\ell} V_{2;\ell}^H & D_{2;4} \end{array}\right] \left[\begin{array}{c||c|c} V_{2;\ell}' & V_{2;\ell} & 0 \\ \hline 0 & 0 & I \end{array}\right] \\ = \left[\begin{array}{c||c|c} U_{1;r} V_{1;r,3}^H V_{2;\ell}' & U_{1;r} V_{1;r,3}^H V_{2;\ell} & U_{1;r} V_{1;r,4}^H \\ \hline\hline D_{2;3}(I - V_{2;\ell} V_{2;\ell}^H) V_{2;\ell}' & D_{2;3} V_{2;\ell} & U_{2;r} V_{2;r}^H \\ \hline 0 & U_{2;\ell} & D_{2;4} \end{array}\right]. \end{array} \tag{5.50}$$

Again, $V'_{2,\ell}$ can be taken out to the second member. Notice also that the size of the system has increased slightly – this is a convenience which keeps the entries in the main block diagonal square, an alternative strategy will be briefly discussed later. We have now reached the 'bottom' as far as this latter part of the matrix is concerned. We can now start eliminating the newly created spurious entries which are all 'skinny' and share column bases. At the bottom we find

$$\left[\begin{array}{c|c||c} U_{\ell,4} & U_{2;\ell} & D_{2;4} \end{array}\right]. \tag{5.51}$$

This can now be reduced by a direct column absorption procedure, dual of the procedure presented in theorem 2. Since we have assumed row-independence, the theorem is directly applicable. This step will reduce the matrix to the form

$$\left[\begin{array}{c||c} 0 & D'_{2;4} \end{array}\right] \tag{5.52}$$

in which $D'_{2;4}$ has become square non-singular, and the row basis of the submatrix on top have been modified (without modification of the column basis), and this by 'skinny' calculations. Notice that only submatrices belonging to low-rank off-diagonal blocks are affected by this step, in hierarchical order, as no central matrix resides on top. The adjustment computation can also be restricted to the lowest level in the hierarchy as all the higher levels will follow suit. Since this is an important ingredient in preserving low computational complexity, we make this step more explicit. The full submatrix to be reduced has the following form:

$$\left[\begin{array}{c||c||c|c} D_1 V_\ell & U_{1;r} V_{1;r,3}^H & U_{1;r} V_{1;r,3}^H V_{2;\ell} & U_{1;r} V_{1;r,4}^H \\ \hline\hline U_{\ell,3} & D_{2;3}(I - V_{2;\ell} V_{2;\ell}^H) & D_{2;3} V_{2;\ell} & U_{2;r} V_{2;r}^H \\ \hline U_{\ell,4} & 0 & U_{2;\ell} & D_{2;4} \end{array}\right]. \tag{5.53}$$

The bottom block rows in this expression have the necessary HSS compatibility relationship to allow for HSS column absorption. $D_1 V_\ell$ involves new data resulting from eliminating the bottom block. The computation of this term is unavoidable as this is the only place in the matrix where the data survives, but it has already been executed as part of the procedure to computer $D_1 - D_1 V_\ell V_\ell^H$ described earlier in this section. The entries in the first block-column all have the same reduced row vector $[I_{\delta_1}]$. The same is true one level down (where a similar procedure has been performed), $D_{2;3} V_{2;\ell}$ is what remains from that elimination procedure. The overall row basis matrix that will be affected by the present HSS column absorption procedure has now the form

$$\left[\begin{array}{c||c|c} I_{\delta_1} & & \\ \hline & I_{\delta_2} & \\ \hline & & V_{2;r}^H \end{array}\right] \tag{5.54}$$

where it may be remarked that the columns on top of the zero entries will be unaffected by the absorption procedure. The procedure scrambles this vector to produce a new row basis, and after reordering of columns produces the right sub matrix

$$\left[\begin{array}{c||c|c} U_{1;r}V_{1;r,3}^H & [D_1V_\ell \;\; U_{1;r}V_{1;r,3}^H V_{2;\ell}]V_{2;r.4}^{\prime,H} & U_{1;r}V_{1;r,5}^{\prime H} \\ \hline\hline D_{2;3}(I - V_{2;\ell}V_{2;\ell}^H) & [U_{\ell.3} \;\; D_{2;3}V_{2;\ell}]V_{2;r,4}^{\prime H} & U_{2;r}V_{2;r,5}^{\prime H} \\ \hline 0 & 0 & D_{2;4}^{\prime} \end{array}\right]. \tag{5.55}$$

It should be clear that the HSS relations still hold (although the rank has necessarily doubled) both for the row bases and for the column bases. This procedure can now be repeated one step further (concerning the rows of hierarchical index 2,3), involving an absorption of $[U_{\ell 3} \;\; D_{2;3}V_{2;\ell}]V_{2;r,4}^{\prime H}$ into $D_{2;3}(I - V_{2;\ell}V_{2;\ell}^H)$, a procedure that is again of the HSS absorption type and can be executed at an even lower level (level 3). The result so far yields the following general form (redefining entries and dropping primes for better visualization)

$$\left[\begin{array}{c|c||c||c|c} D_{2;1} & U_{2;r,1}V_{2;r,2}^H & U_{1;r,1}V_{2;r,3}^H & U_{1;r,1}V_{1;r,4}^H & U_{1;r,1}V_{1;r,5}^H \\ \hline U_{2;\ell,2}V_{2;\ell,1}^H & D_{2;2} & U_{1;r,2}V_{2;r,3}^H & U_{1;r,2}V_{1;r,4}^H & U_{1;r,2}V_{1;r,5}^H \\ \hline\hline 0 & 0 & 0 & D_{2;4} & U_{2;r,3}V_{2;r,5}^H \\ \hline 0 & 0 & 0 & 0 & D_{2;5} \end{array}\right] \tag{5.56}$$

in which all HSS relations hold for as far as applicable. To proceed to the levels with indices 2;1 and 2;2, we must revert to the 1;1 matrix computed earlier (which was the original $D_1(I - V_\ell V_\ell^H)$, and which, as we have shown, still has the HSS form, as shown schematically in the previous equation). This matrix must now undergo the same second level treatment as was done on the original D_2 block, with similar result. The procedure entails the HSS absorption of the third block column in the previous equation into the first two blocks, As this procedure does not affect the fourth and fifth block column, the end result will have the form (again dropping primes)

$$\left[\begin{array}{c|c||c||c|c} D_{2;1} & U_{2;r,1}V_{2;r,2}^H & 0 & U_{1;r,1}V_{1;r,4}^H & U_{1;r,1}V_{1;r,5}^H \\ \hline 0 & D_{2;2} & 0 & U_{1;r,2}V_{1;r,4}^H & U_{1;r,2}V_{1;r,5}^H \\ \hline\hline 0 & 0 & 0 & D_{2;4} & U_{2;r,3}V_{2;r,5}^H \\ \hline 0 & 0 & 0 & 0 & D_{2;5} \end{array}\right] \tag{5.57}$$

where, again, HSS relations have been preserved wherever applicable. If the original system was indeed row independent, then the diagonals blocks in this expression will be square invertible, and a QR factorization of the original system of equations has been achieved, which can simply be solved by an HSS-type back substitution that can be efficiently executed (i.e., by visiting all algebraically relevant data only once, using the HSS relationships). In case the original matrix is not row independent, then a further row reduction can be done on the matrix obtained so far, using a simplified form of the algorithm derived so far.

The dual form of the method presented so far merits some attention. Suppose that our original system is not row- but column independent (i.e., it is a tall system). Is a reduction procedure as used in the previous algorithm that preserves the Moore-Penrose property possible? We show that this is indeed the case.

Proposition 3. *Let*

$$\left[\begin{array}{c|c} D_1 & U_u V_u^H \\ \hline U_\ell V_\ell^H & D_2 \end{array}\right] \tag{5.58}$$

be given as the top-level of a column independent HSS representation in which the matrices $D_i, i = 1, 2$ are HSS compatible, and in which U_u is isometric. Then an equivalent Moore-Penrose system is given by

$$\left[\begin{array}{c|c} (I - U_u U_u^H) D_1 & 0 \\ \hline U_u^H D_1 & V_u^H \\ \hline U_\ell V_\ell^H & D_2 \end{array}\right]. \tag{5.59}$$

Proof. The proof follows from the following two observations:

1. Let U_u' be an isometric matrix whose columns span the orthogonal complement of the (column) range of U_u. Then
$$U_u'^H D_1 = U_u'^H (I - U_u U_u^H) D_1; \tag{5.60}$$
2. if V^H is an isometric matrix then the system $V^H Ax = b$ is equivalent to the system $Ay = Vb$ in the sense that if y solves the Moore-Penrose problem of the latter, then x=y is the Moore-Penrose solution of the former.

The required transformation now follows from the identity

$$\left[\begin{array}{c|c} U_u'^H & \\ \hline U_u^H & \\ \hline & I \end{array}\right] \left[\begin{array}{c|c} D_1 & U_u V_u^H \\ \hline U_\ell V_\ell^H & D_2 \end{array}\right] = \left[\begin{array}{c|c} U_u'^H D_1 & 0 \\ \hline U_u^H D_1 & V_u^H \\ \hline U_\ell V_\ell^H & D_2 \end{array}\right]. \tag{5.61}$$

An application of the properties mentioned above allows one to eliminate $U_u'^H$ so that the modification of the entries only involves the product $U_u^H D_1$, which can efficiently be relegated to the lower hierarchical level. □

Notice that just as before the HSS relations remain in place. We can summarize the procedures in the following conceptual theorem. In the formulation 'URV-type' means: using orthogonal transformation on the rows (for the U-factor) and the columns (for the V factor).

Theorem 1. *A matrix represented in HSS form can be constructively reduced to a non-singular block upper matrix in HSS form using efficient, HSS compatible transformations of the URV type.*

6. Discussion and conclusions

A number of observations seem relevant:

1. the HSS form of an invertible matrix is stable under inversion, it is not too hard to prove that the inverse of such a matrix has an HSS representation of the same complexity as the original. However, an efficient algorithm to compute it has not been presented yet to our knowledge. The present paper goes some way towards this end, it shows at least that the URV form can be

computed efficiently. The result even extends to non-invertible matrices and their Moore-Penrose inverse;
2. the theoretical results presented in this paper have not been tested numerically. This will be done in the near future;
3. as mentioned in the introduction, the HSS form is not as generally available as one would wish. 1D and 2D scattering problems can be brought into the form through the use of multipole theory. For 3D problems, the reduction is far from evident. Here also we are still lacking a good reduction theory. The same can be said about finite element systems: for 1D and 2D cases there is a forthcoming reduction possible as is indicated in the thesis of T. Pals [11];
4. another interesting but as yet unsolved problem is the determination of (close to optimal) preconditioners in HSS form; rank reductions in the HSS form would amount to multi-resolution approximations;
5. the representation for HSS forms discussed in the introduction amounts to a 'system theory on a tree' much as is the case for the multi-resolution theory of Alpay and Volok [1]. However, the representations are fundamentally different: in our case the system theory represents *computational states*, i.e., intermediate data as they are stored in a hierarchical computation, while in the Alpay-Volok case, the states parallel the multi-resolution, a higher up state consists of a summary or average of lower lying states. Although the states have therefore very different semantics, from a system theoretical point of view they are states indexed by a tree and hence operators acting on these states will have similar effects. In particular, the shift operator and its adjoint as well as the level exchange operator presented in the introductory section have the same meaning in both theories (the definition of these operators differ somewhat due to different normalizations). A major difference, however, is that in our case the state space structure is not uniform, while in the Alpay-Volok it is, so that the system in their case can be advantageously reduced to a sequentially semi-separable form.

Acknowledgements

The authors wish to thank a number of colleagues for discussions and interactions with them on the general theme of HSS systems, to wit Alle-Jan van der Veen, Ming Gu, Timoghy Pals, W. Lyons and Zhifeng Sheng.

References

[1] D. Alpay and D. Volok. Point evaluation and hardy space on a homogeneous tree. *Integral Equations and Operator Theory*, vol 53 (2005), pp. 1–22.

[2] S. Chandrasekaran, P. Dewilde, M. Gu, W. Lyons, and T. Pals. A fast solver for hss representations via sparse matrices. In *Technical Report*. Delft University of Technology, August 2005.

[3] S. Chandrasekaran, M. Gu, and T. Pals. A fast and stable solver for smooth recursively semi-separable systems. In *SIAM Annual Conference, San Diego and SIAM Conference of Linear Algebra in Controls, Signals and Systems, Boston*, 2001.

[4] S. Chandrasekaran, M. Gu, and T. Pals. Fast and stable algorithms for hierarchically semi-separable representations. In *Technical Report.* University of California at Santa Barbara, April 2004.

[5] P. Dewilde, K. Diepold, and W. Bamberger. A semi-separable approach to a tridiagonal hierarchy of matrices with application to image flow analysis. In *Proceedings MTNS*, 2004.

[6] P. Dewilde and A.-J. van der Veen. *Time-varying Systems and Computations.* Kluwer, 1998.

[7] P. Dewilde and A.-J. van der Veen. Inner-outer factorization and the inversion of locally finite systems of equations. *Linear Algebra and its Applications*, page *to appear*, 2000.

[8] Y. Eidelman and I. Gohberg. On a new class of structured matrices. *Notes distributed at the 1999 AMS-IMS-SIAM Summer Research Conference*, Structured Matrices in Operator Theory, Numerical Analysis, Control, Signal and Image Processing, 1999.

[9] I. Gohberg, T. Kailath, and I. Koltracht. Linear complexity algorithms for semiseparable matrices. *Integral Equations and Operator Theory*, 8:780–804, 1985.

[10] L. Greengard and V. Rokhlin. A fast algorithm for particle simulations. *J. Comp. Phys.*, 73:325–348, 1987.

[11] T. Pals. *Multipole for Scattering Computations: Spectral Discretization, Stabilization, Fast Solvers.* PhD thesis, Department of Electrical and Computer Engineering, University of California, Santa Barbara, 2004.

[12] V. Rokhlin. Applications of volume integrals to the solution of pde's. *J. Comp. Phys.*, 86:414–439, 1990.

[13] A.J. van der Veen. Time-varying lossless systems and the inversion of large structured matrices. *Archiv f. Elektronik u. Übertragungstechnik*, 49(5/6):372–382, September 1995.

Patrick Dewilde
Department of Electrical Engineering
Delft University of Technology
Delft, The Netherlands
e-mail: `p.dewilde@ewi.tudelft.nl`

Shivkumar Chandrasekaran
Department of Electrical and Computer Engineering
University of California
Santa Barbara, CA 93106-9560, USA
e-mail: `shiv@ece.ucsb.edu`

Operator Theory:
Advances and Applications, Vol. 167, 87–126

Methods from Multiscale Theory and Wavelets Applied to Nonlinear Dynamics

Dorin Ervin Dutkay and Palle E.T. Jorgensen

Abstract. We show how fundamental ideas from signal processing, multiscale theory and wavelets may be applied to nonlinear dynamics.

The problems from dynamics include iterated function systems (IFS), dynamical systems based on substitution such as the discrete systems built on rational functions of one complex variable and the corresponding Julia sets, and state spaces of subshifts in symbolic dynamics. Our paper serves to motivate and survey our recent results in this general area. Hence we leave out some proofs, but instead add a number of intuitive ideas which we hope will make the subject more accessible to researchers in operator theory and systems theory.

Mathematics Subject Classification (2000). 42C40, 42A16, 43A65, 42A65.

Keywords. Nonlinear dynamics, martingale, Hilbert space, wavelet.

1. Introduction

In the past twenty years there has been a great amount of interest in the theory of wavelets, motivated by applications to various fields such as signal processing, data compression, tomography, and subdivision algorithms for graphics. (Our latest check on the word "wavelet" in Google turned up over one million and a half results, 1,590,000 to be exact.) It is enough here to mention two outstanding successes of the theory: JPEG 2000, the new standard in image compression, and the WSQ (wavelet scalar quantization) method which is used now by the FBI to store its fingerprint database. As a mathematical subject, wavelet theory has found points of interaction with functional and harmonic analysis, operator theory, ergodic theory and probability, numerical analysis and differential equations. With the explosion of information due to the expansion of the internet, there came a need for faster algorithms and better compression rates. These have motivated the research for new examples of wavelets and new wavelet theories.

Research supported in part by the National Science Foundation.

Recent developments in wavelet analysis have brought together ideas from engineering, from computational mathematics, as well as fundamentals from representation theory. This paper has two aims: One to stress the interconnections, as opposed to one aspect of this in isolation; and secondly to show that the fundamental Hilbert space ideas from the linear theory in fact adapt to a quite wide class of nonlinear problems. This class includes random-walk models based on martingales. As we show, the theory is flexible enough to allow the adaptation of pyramid algorithms to computations in the dynamics of substitution theory as it is used in discrete dynamics; e.g., the complex dynamics of Julia sets, and of substitution dynamics.

Our subject draws on ideas from a variety of directions. Of these directions, we single out quadrature-mirror filters from signal/image processing. High-pass/low-pass signal processing algorithms have now been adopted by pure mathematicians, although they historically first were intended for speech signals, see [55]. Perhaps unexpectedly, essentially the same quadrature relations were rediscovered in operator-algebra theory, and they are now used in relatively painless constructions of varieties of wavelet bases. The connection to signal processing is rarely stressed in the math literature. Yet, the flow of ideas between signal processing and wavelet math is a success story that deserves to be told. Thus, mathematicians have borrowed from engineers; and the engineers may be happy to know that what they do is used in mathematics.

Our presentation serves simultaneously to motivate and to survey a number of recent results in this general area. Hence we leave out some proofs, but instead we add a number of intuitive ideas which we hope will make the subject more accessible to researchers in operator theory and systems theory. Our theorems with full proofs will appear elsewhere. An independent aim of our paper is to point out several new and open problems in nonlinear dynamics which could likely be attacked with the general multiscale methods that are the focus of our paper.

The first section presents background material from signal processing and from wavelets in a form which we hope will be accessible to operator theorists, and to researchers in systems theory. This is followed by a presentation of some motivating examples from nonlinear dynamics. They are presented such as to allow the construction of appropriate Hilbert spaces which encode the multiscale structure. Starting with a state space X from dynamics, our main tool in the construction of a multiscale Hilbert space $H(X)$ is the theory of random walk and martingales. The main section of our paper serves to present our recent new and joint results.

2. Connection to signal processing and wavelets

We will use the term "filter" in the sense of signal processing. In the simplest case, time is discrete, and a signal is just a numerical sequence. We may be acting on a space of signals (sequences) using the operation of Cauchy product; and the operations of down-sampling and up-sampling. This viewpoint is standard in the engineering literature, and is reviewed in [19] (see also [10] and [63]) for the benefit of mathematicians.

A numerical sequence (a_k) represents a filter, but it is convenient, at the same time, also to work with the frequency-response function. By this we mean simply the Fourier series (see (3.3) below) corresponding to the sequence (a_k). This Fourier series is called the filter function, and in one variable we view it as a function on the one-torus. (This is using the usual identification of periodic functions on the line with functions on the torus.) The advantage of this dual approach is that Cauchy product of sequences then becomes pointwise product of functions.

We will have occasion to also work with several dimensions d, and then our filter function represents a function on the d-torus. While signal processing algorithms are old, their use in wavelets is of more recent vintage. From the theory of wavelets, we learn that filters are used in the first step of a wavelet construction, the so-called multiresolution approach. In this approach, the problem is to construct a function on $\mathbb{R}$ or on $\mathbb{R}^d$ which satisfies a certain renormalization identity, called the scaling equation, and we recall this equation in two versions (2.2) and (3.2) below. The numbers (a_k) entering into the scaling equation turn out to be the very same numbers the signal processing engineers discovered as *quadrature-mirror filters*.

A class of wavelet bases are determined by filter functions. We establish a duality structure for the category of filter functions and classes of representations of a certain C^*-algebra. In the process, we find new representations, and we prove a correspondence which serves as a computational device. Computations are done in the sequence space ℓ^2, while wavelet functions are in the Hilbert space $L^2(\mathbb{R})$, or some other space of functions on a continuum. The trick is that we choose a subspace of $L^2(\mathbb{R})$ in which computations are done. The choice of subspace is dictated by practical issues such as resolution in an image, or refinement of frequency bands.

We consider non-abelian algebras containing canonical maximal abelian subalgebras, i.e., $C(X)$ where X is the Gelfand space, and the representations define measures μ on X. Moreover, in the examples we study, it turns out that X is an affine iterated function system (IFS), of course depending on the representation. In the standard wavelet case, X may be taken to be the unit interval. Following Wickerhauser et al. [30], these measures μ are used in the characterization and analysis of wavelet packets.

Orthogonal wavelets, or wavelet frames, for $L^2\left(\mathbb{R}^d\right)$ are associated with quadrature-mirror filters (QMF), a set of complex numbers which relate the dyadic scaling of functions on $\mathbb{R}^d$ to the $\mathbb{Z}^d$-translates. In the paper [53], we show that generically, the data in the QMF-systems of wavelets are minimal, in the sense that it cannot be nontrivially reduced. The minimality property is given a geometric formulation in $\ell^2\left(\mathbb{Z}^d\right)$; and it is then shown that minimality corresponds to irreducibility of a wavelet representation of the Cuntz algebra $\mathcal{O}_N$. Our result is that this family of representations of $\mathcal{O}_N$ on $\ell^2\left(\mathbb{Z}^d\right)$ is irreducible for a generic set of values of the parameters which label the wavelet representations. Since MRA-wavelets correspond to representations of the Cuntz algebras $\mathcal{O}_N$, we then get, as a bonus, results about these representations.

Definition 2.1. (Ruelle's operators.) Let (X, μ) be a finite Borel measure space, and let $r\colon X \to X$ be a finite-to-one mapping of X onto itself. (The measure μ will be introduced in Section 4 below, and it will be designed so as to have a certain strong invariance property which generalizes the familiar property of Haar measure in the context of compact groups.) Let $V\colon X \to [0, \infty)$ be a given measurable function. We then define an associated operator $R = R_V$, depending on both V and the endomorphism r, by the following formula

$$R_V f(x) = \frac{1}{\# r^{-1}(x)} \sum_{r(y)=x} V(y) f(y), \qquad f \in L^1(X, \mu). \tag{2.1}$$

Each of the operators R_V will be referred to as a Ruelle operator, or a transition operator; and each R_V clearly maps positive functions to positive functions. (When we say "positive" we do not mean "strictly positive", but merely "non-negative".)

We refer to our published papers/monographs [54, 40, 45] about the spectral picture of a positive transition operator R (also called the Ruelle operator, or the Perron–Frobenius–Ruelle operator). Continuing [18], it is shown in [54] that a general family of Hilbert-space constructions, which includes the context of wavelets, may be understood from the so-called Perron–Frobenius eigenspace of R. This eigenspace is studied further in the book [19] by Jorgensen and O. Bratteli; see also [55]. It is shown in [54] that a *scaling identity* (alias *refinement equation*)

$$\varphi(x) = \sqrt{N} \sum_{k \in \mathbb{Z}} a_k \varphi(Nx - k) \tag{2.2}$$

may be viewed generally and abstractly. Variations of (2.2) open up for a variety of new applications which we outline briefly below.

3. Motivating examples, nonlinearity

In this paper we outline how basic operator-theoretic notions from wavelet theory (such as the successful multiresolution construction) may be adapted to certain state spaces in nonlinear dynamics. We will survey and extend our recent papers on multiresolution analysis of state spaces in symbolic shift dynamics $X(A)$, and on a class of substitution systems $X(r)$ which includes the Julia sets; see, e.g., our papers [42, 43, 44, 45]. Our analysis of these systems X starts with consideration of Hilbert spaces of functions on X. But already this point demands new tools. So the first step in our work amounts to building the right Hilbert spaces. That part relies on tools from both operator algebras, and from probability theory (e.g., path-space measures and martingales).

First we must identify the appropriate measures on X, and secondly, we must build Hilbert spaces on associated extension systems X_∞, called generalized solenoids.

The appropriate measures μ on X will be constructed using David Ruelle's thermodynamical formalism [71]: We will select our distinguished measures μ on X to have minimal free energy relative to a certain function W on X. The relationship between the measures μ on X and a class of induced measures on the extensions X_∞ is based on a random-walk model which we developed in [44], such that the relevant measures on X_∞ are constructed as path-space measures on paths which originate on X. The transition on paths is governed in turn by a given and prescribed function W on X.

In the special case of traditional wavelets, W is the absolute square of a so-called low-pass wavelet filter. In these special cases, the wavelet-filter functions represent a certain system of functions on the circle $\mathbb{T} = \mathbb{R}/\mathbb{Z}$ which we outline below; see also our paper [45]. Even our analysis in [45] includes wavelet bases on affine Cantor sets in $\mathbb{R}^d$. Specializing our Julia-set cases $X = X(r)$ to the standard wavelet systems, we get $X = \mathbb{T}$, and the familiar approach to wavelet systems becomes a special case.

According to its original definition, a *wavelet* is a function $\psi \in L^2(\mathbb{R})$ that generates an orthonormal basis under the action of two unitary operators: the dilation and the translation. In particular,

$$\{\, U^j T^k \psi \mid j, k \in \mathbb{Z} \,\}$$

must be an orthonormal basis for $L^2(\mathbb{R})$, where

$$Uf(x) = \frac{1}{\sqrt{2}} f\left(\frac{x}{2}\right), \quad Tf(x) = f(x-1), \qquad f \in L^2(\mathbb{R}), x \in \mathbb{R}. \tag{3.1}$$

One of the effective ways of getting concrete wavelets is to first look for a *multiresolution analysis* (*MRA*), that is to first identify a suitable telescoping nest of subspaces $(\mathcal{V}_n)_{n\in\mathbb{Z}}$ of $L^2(\mathbb{R})$ that have trivial intersection and dense union, $\mathcal{V}_n = U\mathcal{V}_{n-1}$, and $\mathcal{V}_0$ contains a *scaling function* φ such that the translates of φ form an orthonormal basis for $\mathcal{V}_0$.

The scaling function will necessarily satisfy an equation of the form

$$U\varphi = \sum_{k\in\mathbb{Z}} a_k T^k \varphi, \tag{3.2}$$

called the scaling equation (see also (2.2)).

To construct wavelets, one has to choose a *low-pass filter*

$$m_0(z) = \sum_{k\in\mathbb{Z}} a_k z^k \tag{3.3}$$

and from this obtain the scaling function φ as the fixed point of a certain cascade operator. The wavelet ψ (mother function) is constructed from φ (a father function, or scaling function) with the aid of two closed subspaces $\mathcal{V}_0$, and $\mathcal{W}_0$. The scaling function φ is obtained as the solution to a scaling equation (3.2), and its integral translates generates $\mathcal{V}_0$ (the initial resolution), and ψ with its integral translates generating $\mathcal{W}_0 := \mathcal{V}_1 \ominus \mathcal{V}_0$, where $\mathcal{V}_1$ is the next finer resolution subspace containing the initial resolution space $\mathcal{V}_0$, and obtained from a one-step refinement of $\mathcal{V}_0$.

Thus the main part of the construction involves a clever choice for the low-pass filter m_0 that gives rise to nice orthogonal wavelets.

MRAs in geometry and operator theory

Let X be a compact Hausdorff space, and let $r\colon X \to X$ be a finite-to-one continuous endomorphism, mapping X onto itself. As an example, $r = r(z)$ could be a rational mapping of the Riemann sphere, and X could be the corresponding Julia set; or X could be the state space of a subshift associated to a 0-1 matrix.

Due to work of Brolin [25] and Ruelle [72], it is known that for these examples, X carries a unique maximal entropy, or minimal free-energy measure μ also called strongly r-invariant; see Lemma 4.3 (ii), and equations (4.20) and (4.27)–(4.28) below. For each point $x \in X$, the measure μ distributes the "mass" equally on the finite number of solutions y to $r(y) = x$.

We show that this structure in fact admits a rich family of wavelet bases. Now this will be on a Hilbert space which corresponds to $L^2(\mathbb{R})$ in the familiar case of multiresolution wavelets. These are the wavelets corresponding to scaling x to Nx, by a fixed integer N, $N \geq 2$. In that case, $X = \mathbb{T}$, the circle in $\mathbb{C}$, and $r(z) = z^N$. So even in the "classical" case, there is a "unitary dilation" from $L^2(X)$ to $L^2(\mathbb{R})$ in which the Haar measure on $\mathbb{T}$ "turns into" the Lebesgue measure on $\mathbb{R}$.

Our work so far, on examples, shows that this viewpoint holds promise for understanding the harmonic analysis of Julia sets, and of other iteration systems. In these cases, the analogue of $L^2(\mathbb{R})$ involves quasi-invariant measures μ_∞ on a space X_∞, built from X in a way that follows closely the analogue of passing from $\mathbb{T}$ to $\mathbb{R}$ in wavelet theory. But μ_∞ is now a path-space measure. The translations by the integers $\mathbb{Z}$ on $L^2(\mathbb{R})$ are multiplications in the Fourier dual. For the Julia examples, this corresponds to the coordinate multiplication, i.e., multiplication by z on $L^2(X, \mu)$, a normal operator. We get a corresponding covariant system on X_∞, where multiplication by $f(z)$ is unitarily equivalent to multiplication by $f(r(z))$. But this is now on the Hilbert space $L^2(X_\infty)$, and defined from the path-space measure μ_∞. Hence all the issues that are addressed since the mid-1980's for $L^2(\mathbb{R})$-wavelets have analogues in the wider context, and they appear to reveal interesting spectral theory for Julia sets.

3.1. Spectrum and geometry: wavelets, tight frames, and Hilbert spaces on Julia sets

3.1.1. Background. Problems in dynamics are in general irregular and chaotic, lacking the internal structure and homogeneity which is typically related to group actions. Lacking are the structures that lie at the root of harmonic analysis, i.e., the torus or $\mathbb{R}^d$.

Chaotic attractors can be viewed simply as sets, and they are often lacking the structure of manifolds, or groups, or even homogeneous manifolds. Therefore, the "natural" generalization, and question to ask, then depends on the point of view, or on the application at hand.

Hence, many applied problems are typically not confined to groups; examples are turbulence, iterative dynamical systems, and irregular sampling. But there are many more. Yet harmonic analysis and more traditional MRA theory begins with $\mathbb{R}^d$, or with some such group context.

The geometries arising in wider contexts of applied mathematics might be attractors in discrete dynamics, in substitution tiling problems, or in complex substitution schemes, of which the Julia sets are the simplest. These geometries are not tied to groups at all. And yet they are very algorithmic, and they invite spectral-theoretic computations.

Julia sets are prototypical examples of chaotic attractors for iterative discrete dynamical systems; i.e., for iterated substitution of a fixed rational function $r(z)$, for z in the Riemann sphere. So the Julia sets serve beautifully as a testing ground for discrete algorithms, and for analysis on attractors. In our papers [44], [42] we show that multiscale/MRA analysis adapts itself well to discrete iterative systems, such as Julia sets, and state space for subshifts in symbolic dynamics. But so far there is very little computational harmonic analysis outside the more traditional context of $\mathbb{R}^d$.

Definition 3.1. Let S be the Riemann sphere, and let $r\colon S \to S$ be a rational map of degree greater than one. Let r^n be the nth iteration of r, i.e., the nth iterated substitution of r into itself. The Fatou set $F(r)$ of r is the largest open set U in S such that the sequence r^n, restricted to U, is a normal family (in the sense of Montel). The *Julia set* $X(r)$ is the complement of $F(r)$.

Moreover, $X(r)$ is known to behave (roughly) like an attractor for the discrete dynamical system r^n. But it is more complicated: the Julia set $X(r)$ is the locus of expanding and chaotic behavior; e.g., $X(r)$ is equal to the closure of the set of repelling periodic points for the dynamical system r^n in S.

In addition, $X = X(r)$ is the minimal closed set X, such that $|X| > 2$, and X is invariant under the branches of r^{-1}, the inverses of r.

While our prior research on Julia sets is only in the initial stages, there is already some progress. We are especially pleased that Hilbert spaces of martingales have offered the essential framework for analysis on Julia sets. The reason is that martingale tools are more adapted to irregular structures than are classical Fourier methods.

To initiate a harmonic analysis outside the classical context of the tori and of $\mathbb{R}^d$, it is thus natural to begin with the Julia sets. Furthermore, the Julia sets are already of active current research interest in geometry and dynamical systems.

But so far, there are only sporadic attempts at a harmonic analysis for Julia sets, let alone the more general geometries outside the classical context of groups. Our prior work should prepare us well for this new project. Now, for Julia-set geometries, we must begin with the Hilbert space. And even that demands new ideas and new tools.

To accommodate wavelet solutions in the wider context, we built new Hilbert spaces with the use of tools from martingale theory. This is tailored to applications we have in mind in geometry and dynamical systems involving Julia sets, fractals, subshifts, or even more general discrete dynamical systems.

The construction of solutions requires a combination of tools that are somewhat non-traditional in the subject.

A number of wavelet constructions in pure and applied mathematics have a common operator-theoretic underpinning. It may be illustrated with the following operator system: $\mathcal{H}$ some Hilbert space; $U\colon \mathcal{H} \to \mathcal{H}$ a unitary operator; $V\colon \mathbb{T} \to \mathcal{U}(\mathcal{H})$ a unitary representation. Here $\mathbb{T}$ is the 1-torus, $\mathbb{T} = \mathbb{R}/\mathbb{Z}$, and $\mathcal{U}(\mathcal{H})$ is the group of unitary operators in $\mathcal{H}$.

The operator system satisfies the identity

$$V(z)^{-1} UV(z) = zU, \qquad z \in \mathbb{T}. \tag{3.4}$$

Definition 3.2. We say that U is homogeneous of degree one with respect to the scaling automorphisms defined by $\{ V(z) \mid z \in \mathbb{T} \}$ if (3.4) holds. In addition, U must have the spectral type defined by Haar measure. We say that U is homogeneous of degree one if it satisfies (3.4) for some representation $V(z)$ of $\mathbb{T}$.

In the case of the standard wavelets with scale N, $\mathcal{H} = L^2(\mathbb{R})$, where in this case the two operators U and T are the usual N-adic scaling and integral translation operators; see (3.1) above giving U for the case $N = 2$.

If $\mathcal{V}_0 \subset \mathcal{H}$ is a resolution subspace, then $\mathcal{V}_0 \subset U^{-1}\mathcal{V}_0$. Set $\mathcal{W}_0 := U^{-1}\mathcal{V}_0 \ominus \mathcal{V}_0$, $Q_0 :=$ the projection onto $\mathcal{W}_0$, $Q_k = U^{-k} Q_0 U^k$, $k \in \mathbb{Z}$, and

$$V(z) = \sum_{k=-\infty}^{\infty} z^k Q_k, \qquad z \in \mathbb{T}. \tag{3.5}$$

Then it is easy to check that the pair $(U, V(z))$ satisfies the commutation identity (3.4).

Let $(\psi_i)_{i \in I}$ be a Parseval frame in $\mathcal{W}_0$. Then

$$\left\{ U^k T_n \psi_i \;\middle|\; i \in I,\ k, n \in \mathbb{Z} \right\} \tag{3.6}$$

is a Parseval frame for $\mathcal{H}$. (Recall, a Parseval frame is also called a normalized tight frame.)

Turning the picture around, we may start with a system $(U, V(z))$ which satisfies (3.4), and then reconstruct wavelet (or frame) bases in the form (3.6). To do this in the abstract, we must first understand the multiplicity function calculated for $\{ T_n \mid n \in \mathbb{Z} \}$ when restricted to the two subspaces $\mathcal{V}_0$ and $\mathcal{W}_0$. But such a multiplicity function may be be defined for any representation of an abelian C^*-algebra acting on $\mathcal{H}$ which commutes with some abstract scaling automorphism $(V(z))$.

This technique can be used for the Hilbert spaces associated to Julia sets, to construct wavelet (and frame) bases in this context.

3.1.2. Wavelet filters in nonlinear models. These systems are studied both in the theory of symbolic dynamics, and in C^*-algebra theory, i.e., for the Cuntz–Krieger algebras.

It is known [25, 71] that, for these examples, X carries a unique maximal entropy, or minimal free-energy measure μ. The most general case when (X, r) admits a strongly r-invariant measure is not well understood.

The intuitive property of such measures μ is this: For each point x in X, μ distributes the "mass" equally on the finite number of solutions y to $r(y) = x$. Then μ is obtained by an iteration of this procedure, i.e., by considering successively the finite sets of solutions y to $r^n(y) = x$, for all $n = 1, 2, \ldots$; taking an average, and then limit. While the procedure seems natural enough, the structure of the convex set $K(X, r)$ of all the r-invariant measures is worthy of further study. For the case when $X = \mathbb{T}$ the circle, μ is unique, and it is the Haar measure on $\mathbb{T}$.

The invariant measures are of interest in several areas: in representation theory, in dynamics, in operator theory, and in C^*-algebra theory. Recent work of Dutkay Jorgensen [43, 44, 42] focuses on special convex subsets of $K(X, r)$ and their extreme points. This work in turn is motivated by our desire to construct wavelet bases on Julia sets.

Our work on $K(X, r)$ extends what is know in traditional wavelet analysis. While linear wavelet analysis is based on Lebesgue measure on $\mathbb{R}^d$, and Haar measure on the d-torus $\mathbb{T}^d$, the nonlinear theory is different, and it begins with results from geometric measure theory: We rely on results from [25], [60], [50], and [61]. And our adaptation of spectral theory in turn is motivated by work by Baggett et al. in [8, 7].

Related work reported in [61] further ties in with exciting advances on graph C^*-algebras pioneered by Paul Muhly, Iain Raeburn and many more in the C^*-algebra community; see [8], [10], [14], [20], [21], [31], [38] [65], [66], [68].

In this section, we establish specific basis constructions in the context of Julia sets, function systems, and self-similar operators.

Specifically, we outline a construction of a multiresolution/wavelet analysis for the Julia set X of a given rational function $r(z)$ of one complex variable, i.e., $X = \text{Julia}(r)$. Such an analysis could likely be accomplished with some kind of wavelet representation induced by a normal operator T which is unitarily equivalent to a function of itself. Specifically, T is unitarily equivalent to $r(T)$, i.e.,

$$UT = r(T)U, \qquad \text{for some unitary operator } U \text{ in a Hilbert space } \mathcal{H}. \tag{3.7}$$

Even the existence of these representations would seem to be new and significant in operator theory.

There are consequences and applications of such a construction: First we will get the existence of a finite set of some special generating functions m_i on the Julia set X that may be useful for other purposes, and which are rather difficult to construct by alternative means. The relations we require for a system

$m_0, \ldots, m_{N-1}$ of functions on $X(r)$ are as follows.

$$\frac{1}{N} \sum_{\substack{y \in X(r) \\ r(y)=x}} \overline{m_i(y)}\, m_j(y) h(y) = \delta_{i,j} h(x) \qquad \text{for a.e. } x \in X(r), \tag{3.8}$$

where h is a Perron–Frobenius eigenfunction for a Ruelle operator R defined on $L^\infty(X(r))$. Secondly, it follows that for each Julia set X, there is an infinite-dimensional group which acts transitively on these systems of functions. The generating functions on X we are constructing are analogous to the more familiar functions on the circle $\mathbb{T}$ which define systems of filters in wavelet analysis. In fact, the familiar construction of a wavelet basis in $\mathcal{H} = L^2(\mathbb{R})$ is a special case of our analysis.

In the standard wavelet case, the rational function r is just a monomial, i.e., $r(z) = z^N$ where N is the number of frequency bands, and X is the circle $\mathbb{T}$ in the complex plane. The simplest choice of the functions $m_0, \ldots, m_{N-1}$ and h in this case is $m_k(z) = z^k$, $z \in \mathbb{T}$, $0 \leq k \leq N-1$, and $h(z) \equiv 1$, $z \in \mathbb{T}$. This represents previous work by many researchers, and also joint work between Jorgensen and Bratteli [19].

Many applied problems are typically not confined to groups; examples are turbulence, iterative dynamical systems, and irregular sampling. But there are many more.

Even though attractors such as Julia sets are highly nonlinear, there are adaptations of the geometric tools from the analysis on $\mathbb{R}$ to the nonlinear setting.

The adaptation of traditional wavelet tools to nonlinear settings begins with an essential step: the construction of an appropriate Hilbert space; hence the martingales.

Jorgensen, Bratteli, and Dutkay have developed a representation-theoretic duality for wavelet filters and the associated wavelets, or tight wavelet frames. It is based on representations of the Cuntz algebra $\mathcal{O}_N$ (where N is the scale number); see [19], [39], [44], [53]. As a by-product of this approach they get infinite families of inequivalent representations of $\mathcal{O}_N$ which are of independent interest in operator-algebra theory.

The mathematical community has long known of operator-theoretic power in Fourier-analytic issues, e.g., Rieffel's incompleteness theorem [68] for Gabor systems violating the Nyquist condition, cf. [34]. We now feel that we can address a number of significant problems combining operator-theoretic and harmonic analysis methods, as well as broadening and deepening the mathematical underpinnings of our subject.

3.2. Multiresolution analysis (MRA)

One reason various geometric notions of multiresolution in Hilbert space (starting with S. Mallat) have proved useful in computational mathematics is that these resolutions are modeled on the fundamental concept of dyadic representation of the real numbers; or more generally on some variant of the classical positional representation of real numbers. In a dyadic representation of real numbers, the shift corresponds to multiplication by 2^{-1}.

The analogue of that in Hilbert space is a unitary operator U which scales functions by 2^{-1}. If the Hilbert space $\mathcal{H}$ is $L^2(\mathbb{R})$, then a resolution is a closed subspace $\mathcal{V}_0$ in $\mathcal{H}$ which is invariant under U with the further property that the restriction of U to $\mathcal{V}_0$ is a shift operator. Positive and negative powers of U then scale $\mathcal{H}$ into a nested resolution system indexed by the integers; and this lets us solve algorithmic problems by simply following the standard rules from number theory.

However, there is no reason at all that use of this philosophy should be restricted to function spaces on the line, or on Euclidean space. Other geometric structures from discrete dynamics admit resolutions as well.

Most MRA, Fourier multiresolution analysis (FMRA) and generalized multiresolution analysis (GMRA) wavelet constructions in pure and applied mathematics have a common operator-theoretic underpinning. Consider the operator system $(\mathcal{H}, U, V(z))$ from (3.4) above.

Proposition 3.3. *Once the system $(\mathcal{H}, U, V)$ is given as in (3.4), a scale of closed subspaces $\mathcal{V}_n$ (called resolution subspaces) may be derived from the spectral subspaces of the representation $\mathbb{T} \ni z \mapsto V(z)$, i.e., subspaces $(\mathcal{V}_n)_{n\in\mathbb{Z}}$ such that $\mathcal{V}_n \subset \mathcal{V}_{n+1}$,*

$$\bigcap \mathcal{V}_n = \{0\}, \qquad \bigcup \mathcal{V}_n \text{ is dense in } \mathcal{H},$$

and $U\mathcal{V}_n \subset \mathcal{V}_{n-1}$.

Conversely, if these spaces $(\mathcal{V}_n)$ are given, then $V(z)$ defined by equation (3.5) can be shown to satisfy (3.4) if

$$Q_0 = \text{the orthogonal projection onto } U^{-1}\mathcal{V}_0 \ominus \mathcal{V}_0,$$

and $Q_n = U^{-n}Q_0U^n$, $n \in \mathbb{Z}$.

As a result we note the following criterion.

Proposition 3.4. *A given unitary operator U in a Hilbert space $\mathcal{H}$ is part of some multiresolution system $(\mathcal{V}_n)$ for $\mathcal{H}$ if and only if U is homogeneous of degree one with respect to some representation $V(z)$ of the one-torus $\mathbb{T}$.*

In application the spaces $\mathcal{V}_n$ represent a grading of the entire Hilbert space $\mathcal{H}$, and we say that U scales between the different levels. In many cases, this operator-theoretic framework serves to represent structures which are similar up to scale, structures that arise for example in the study of *fractional Brownian motion (FBM)*. See, e.g., [59], which offers a Hilbert-space formulation of FBM based on a white-noise stochastic integration.

Moreover, this can be done for the Hilbert space $\mathcal{H} = L^2(X_\infty, \mu_\infty)$ associated with a Julia set X. As a result, we get wavelet bases and Parseval wavelets in this context.

This is relevant for the two mentioned classes of examples, Julia sets $X(r)$, and state spaces $X(A)$ of substitutions or of subshifts. Recall that the translations by the integers $\mathbb{Z}$ on $L^2(\mathbb{R})$ are multiplication in the Fourier dual. For the Julia sets, this corresponds to coordinate multiplication, i.e., multiplication by z on the Hilbert space $L^2(X, \mu)$, consequently a normal operator. The construction will involve Markov processes and martingales.

Thus, many of the concepts related to multiresolutions in $L^2(\mathbb{R})$ have analogues in a more general context; moreover, they seem to exhibit interesting connections to the geometry of Julia sets.

3.2.1. Pyramid algorithms and geometry. Several of the co-authors' recent projects involve some variant or other of the Ruelle transfer operator R, also called the Ruelle–Perron–Frobenius operator. In each application, it arises in a wavelet-like setting. But the settings are different from the familiar $L^2(\mathbb{R})$-wavelet setup: one is for affine fractals, and the other for Julia sets generated from iteration dynamics of rational functions $r(z)$.

Thus, there are two general themes in this work. In rough outline, they are like this:

(1) In the paper [45], Dutkay and Jorgensen construct a new class of wavelets on extended fractals based on Cantor-like iterated function systems, e.g., the middle-third Cantor set. Our Hilbert space in each of our constructions is separable. For the fractals, it is built on Hausdorff measure of the same (fractal) dimension as the IFS-Cantor set in question. In the paper [45], we further introduce an associated Ruelle operator R, and it plays a central role. For our wavelet construction, there is, as is to be expected, a natural (equilibrium) measure ν which satisfies $\nu R = \nu$, i.e., a left Perron–Frobenius eigenvector. It corresponds to the Dirac point-measure on the frequency variable $\omega = 0$ (i.e., low-pass) in the standard $L^2(\mathbb{R})$-wavelet setting. It turns out that our measures ν for the IFS-Cantor sets are given by infinite Riesz products, and they are typically singular and have support $=$ the whole circle $\mathbb{T}$. This part of our research is related to recent, completely independent, work by Benedetto et al. on Riesz products.

(2) A second related research direction is a joint project in progress with Ola Bratteli, where we build wavelets on Hilbert spaces induced by Julia sets of rational functions $r(z)$ of one complex variable. Let r be a rational function of degree at least 2, and let $X(r)$ be the corresponding Julia set. Then there is a family of Ruelle operators indexed by certain potentials, and a corresponding family of measures ν on $X(r)$ which satisfy $\nu R = \nu$, again with the measures ν playing a role somewhat analogous to the Dirac point-mass on the frequency variable $\omega = 0$, for the familiar $L^2(\mathbb{R})$-MRA wavelet construction.

Independently of Dutkay–Jorgensen [45], John Benedetto and his co-authors Erica Bernstein and Ioannis Konstantinidis [12] have developed a new Fourier/infinite-product approach to the very same singular measures that arise in the study in [45]. The motivations and applications are different. In [45] the issue is wavelets on fractals, and in [12], it is time-frequency duality.

In this second class of problems, the measures ν are far less well understood; and yet, if known, they would yield valuable insight into the spectral theory of r-iteration systems and the corresponding Julia sets $X(r)$.

3.3. Julia sets from complex dynamics

Wavelet-like constructions are just under the surface in the following brief sketch of problems. The projects all have to do with the kind of dynamical systems X already mentioned; and Ω is one of the standard probability spaces. The space X could be the circle, or a torus, or a solenoid, or a fractal, or it could be a Julia set defined from a rational map in one complex variable. More generally, consider maps r of X *onto* X which generate some kind of dynamics. The discussion is restricted to the case when for each x in X, the pre-image $r^{-1}(\{x\})$ is assumed finite. Using then a construction of Kolmogorov, it can be seen that these systems X admit useful measures which are obtained by a limit construction and by averaging over the finite sets $r^{-1}(\{x\})$; see, e.g., Jorgensen's new monograph [52] and Proposition 4.18 for more details.

We list three related problems:

(1) **Operator algebras**: A more systematic approach, along these lines, to crossed products with endomorphisms. For example, trying to capture the framework of endomorphism-crossed products introduced earlier by Bost and Connes [16].
(2) **Dynamics**: Generalization to the study of onto-maps, when the number and nature of branches of the inverse is more complicated; i.e., where it varies over X, and where there might be overlaps.
(3) **Geometry**: Use of more detailed geometry and topology of Julia sets, in the study of the representations that come from multiplicity considerations. The Dutkay–Jorgensen paper [45] on fractals is actually a special case of Julia-set analysis. (One of the Julia sets X from the above *is* a Cantor set in the usual sense, e.g., the Julia set of $r(z) = 2z - 1/z$; or $r(z) = z^2 - 3$.)

A main theme will be Hilbert spaces built on Julia sets $X(r)$ built in turn on rational functions $r(z)$ in one complex variable. In the case when r is not a monomial, say of degree $N \geq 2$, we construct N functions m_i, $i = 0, 1, \ldots, N-1$, on $X(r)$ which satisfy a set of quadratic conditions (3.8) analogous to the axioms that define wavelet filters. We have done this, so far, using representation theory, in special cases. But our results are not yet in a form where they can be used in computations. By using results on the Hausdorff dimension of Julia sets, it would also be interesting to identify inequivalent representations, both in the case when the Julia set X is fixed but the functions vary, and in the case when X varies.

Moreover, we get an infinite-dimensional "loop group" $G = G(X, N)$ acting transitively on these sets of functions m_i. The group G consists of measurable functions from X into the group U_N of all unitary N by N complex matrices. In particular, our method yields information about this group. Since X can be geometrically complicated, it is not at all clear how such matrix functions $X \to \mathrm{U}_N(\mathbb{C})$ might be constructed directly.

The group G consists of measurable functions $A\colon X(r) \to \mathrm{U}_N(\mathbb{C})$, and the action of A on an (m_i)-system is as follows: $(m_i) \mapsto (m_i^{(A)})$, where

$$m_i^{(A)}(x) = \sum_j A_{i,j}(r(x))m_j(x), \qquad x \in X(r). \tag{3.9}$$

Even when $N = 2$, the simplest non-monomial examples include the Julia sets of $r(z) = z^2 + c$, and the two functions m_0 and m_1 on $X(c) = \mathrm{Julia}(z^2 + c)$ are not readily available by elementary means. The use of operator algebras, representation theory seems natural for this circle of problems.

4. Main results

Multiresolution/wavelet analysis on Julia sets. We attempt to use equilibrium measures (Brolin [25], Ruelle [71], Mauldin–Urbanski [61]) from discrete dynamical systems to specify generalized low-pass conditions in the Julia-set theory. This would seem to be natural, but there are essential difficulties to overcome. While Ruelle's thermodynamical formalism was developed for transfer operators where the weight function W is strictly positive, the applications we have in mind dictate careful attention to the case when W is not assumed strictly positive. The reason is that for our generalized multiresolution analysis on Julia sets, W will be the absolute square of the low-pass filter.

We begin with one example and a lemma. The example realizes a certain class of state spaces from symbolic dynamics; and the lemma spells out a class of invariant measures which will be needed in our Hilbert-space constructions further into the section. The measures from part (ii) in Lemma 4.3 below are often called strongly invariant measures, or equilibrium measures from Ruelle's thermodynamical formalism, see [71].

Example 4.1. Let $N \in \mathbb{Z}_+$, $N \geq 2$ and let $A = (a_{ij})_{i,j=1}^N$ be an N by N matrix with all $a_{ij} \in \{0,1\}$. Set

$$X(A) := \{\, (x_i) \in \prod_{\mathbb{N}} \{1,\dots,N\} \mid A(x_i, x_{i+1}) = 1 \,\}$$

and let $r = r_A$ be the restriction of the shift to $X(A)$, i.e.,

$$r_A(x_1, x_2, \dots) = (x_2, x_3, \dots), \qquad x = (x_1, x_2, \dots) \in X(A).$$

Lemma 4.2. *Let A be as above. Then*

$$\#\, r_A^{-1}(x) = \#\,\{\, y \in \{1,\dots,N\} \mid A(y, x_1) = 1 \,\}.$$

It follows that $r_A\colon X(A) \to X(A)$ is onto iff A is *irreducible*, i.e., iff for all $j \in \mathbb{Z}_N$, there exists an $i \in \mathbb{Z}_N$ such that $A(i,j) = 1$.

Suppose in addition that A is *aperiodic*, i.e., there exists $p \in \mathbb{Z}_+$ such that $A^p > 0$ on $\mathbb{Z}_N \times \mathbb{Z}_N$.

We have the following lemma.

Lemma 4.3. (D. Ruelle [72, 9]) *Let A be irreducible and aperiodic and let $\phi \in C(X(A))$ be given. Assume that ϕ is a Lipschitz function.*

(i) *Set*

$$(R_\phi f)(x) = \sum_{r_A(y)=x} e^{\phi(y)} f(y), \quad \textit{for } f \in C(X(A)).$$

Then there exist $\lambda_0 > 0$,

$$\lambda_0 = \sup\{\, |\lambda| \mid \lambda \in \operatorname{spec}(R_\phi)\,\},$$

$h \in C(X(A))$ strictly positive and ν a Borel measure on $X(A)$ such that

$$R_\phi h = \lambda_0 h,$$

$$\nu R_\phi = \lambda_0 \nu,$$

and $\nu(h) = 1$. The data are unique.

(ii) *In particular, setting*

$$(R_0 f)(x) = \frac{1}{\# r_A^{-1}(x)} \sum_{r_A(y)=x} f(y),$$

we may take $\lambda_0 = 1$, $h = 1$ and $\nu =: \mu_A$, where μ_A is a probability measure on $X(A)$ satisfying the strong invariance property

$$\int_{X(A)} f \, d\mu_A = \int_{X(A)} \frac{1}{\# r_A^{-1}(x)} \sum_{r_A(y)=x} f(y)\, d\mu_A(x), \qquad f \in L^\infty(X(A)).$$

Our next main results, Theorems 4.4 and Theorem 4.6, are operator-theoretic, and they address the following question: Starting with a state space X for a dynamical system, what are the conditions needed for building a global Hilbert space $H(X)$ and an associated multiscale structure? Our multiscales will be formulated in the context of Definitions 3.2 (or equivalently 3.3) above; that is, we define our multiscale structure in terms of a natural Hilbert space $H(X)$ and a certain unitary operator U on $H(X)$ which implements the multiscales.

After our presentation of the Hilbert-space context, we turn to spectral theory: In our next result, we revisit Baggett's dimension-consistency equation. The theorem of Baggett et al. [8] concerns a condition on a certain pair of multiplicity functions for the existence of a class of wavelet bases for $L^2(\mathbb{R}^d)$. In Theorem 4.8 below, we formulate the corresponding multiplicity functions in the context of nonlinear dynamics, and we generalize the Baggett et al. theorem to the Hilbert spaces $H(X)$ of our corresponding state spaces X.

As preparation for Theorem 4.15 below, we introduce a family of L^2-martingales, and we prove that the Hilbert spaces $H(X)$ are L^2-martingale Hilbert spaces. In this context, we prove the existence of an explicit wavelet transform (Theorem 4.15) for our nonlinear state spaces. But in the martingale context, we must work with dilated state spaces X_∞; and in Proposition 4.18, and Theorems 5.1, 5.2, 5.9, and 5.12, we outline the role of our multiscale structure in connection with certain random-walk measures on X_∞. A second issue concerns extreme measures: in Theorem 5.8 (in the next section) we characterize the extreme points in the convex set of these random-walk measures.

Theorem 4.4. *Let $\mathcal{A}$ be a unital C^*-algebra, α an endomorphism on $\mathcal{A}$, μ a state on $\mathcal{A}$ and, $m_0 \in \mathcal{A}$, such that*

$$\mu(m_0^*\alpha(f)m_0) = \mu(f), \qquad f \in \mathcal{A}. \tag{4.1}$$

Then there exists a Hilbert space H, a representation π of $\mathcal{A}$ on H, U a unitary on H, and a vector $\varphi \in \mathcal{A}$, with the following properties:

$$U\pi(f)U^* = \pi(\alpha(f)), \qquad f \in \mathcal{A}, \tag{4.2}$$

$$\langle \varphi \mid \pi(f)\varphi \rangle = \mu(f), \qquad f \in \mathcal{A}, \tag{4.3}$$

$$U\varphi = \pi(\alpha(1)m_0)\varphi \tag{4.4}$$

$$\overline{\mathrm{span}}\{\, U^{-n}\pi(f)\varphi \mid n \geq 0, f \in \mathcal{A} \,\} = H. \tag{4.5}$$

Moreover, this is unique up to an intertwining isomorphism.

We call (H, U, π, φ) the covariant system associated to μ and m_0.

Corollary 4.5. *Let X be a measure space, $r\colon X \to X$ a measurable, onto map and μ a probability measure on X such that*

$$\int_X f\, d\mu = \int_X \frac{1}{\# r^{-1}(x)} \sum_{r(y)=x} f(y)\, d\mu(x). \tag{4.6}$$

Let $h \in L^1(X)$, $h \geq 0$ such that

$$\frac{1}{\# r^{-1}(x)} \sum_{r(y)=x} |m_0(y)|^2 h(y) = h(x), \qquad x \in X.$$

Then there exists (uniquely up to isomorphism) a Hilbert space H, a unitary U, a representation π of $L^\infty(X)$ and a vector $\varphi \in H$ such that

$$U\pi(f)U^{-1} = \pi(f \circ r), \qquad f \in L^\infty(X),$$

$$\langle \varphi \mid \pi(f)\varphi \rangle = \int_X fh\, d\mu, \qquad f \in L^\infty(X),$$

$$U\varphi = \pi(m_0)\varphi,$$

$$\overline{\mathrm{span}}\{\, U^{-n}\pi(f)\varphi \mid n \geq 0, f \in L^\infty(X) \,\} = H.$$

We call (H, U, π, φ) the covariant system associated to m_0 and h.

The Baumslag–Solitar group. Our next theorem is motivated by the theory of representations of a certain group which is called the Baumslag–Solitar group, and which arises by the following simple semidirect product construction: the Baumslag–Solitar group $BS(1, N)$ is the group with two generators u and t and one relation $utu^{-1} = t^N$, where N is a fixed positive integer. Therefore the two operators U of N-adic dilation and T of integral translation operators on $\mathbb{R}$ give a representation of this group. But so do all familiar MRA-wavelet constructions.

Representations of the Baumslag–Solitar group that admit wavelet bases can be constructed also on some other spaces, such as $L^2(\mathbb{R}) \oplus \cdots \oplus L^2(\mathbb{R})$ [13, 41] or some fractal spaces [45].

Hence wavelet theory fits naturally into this abstract setting of special representations of the group $BS(1, N)$ realized on some Hilbert space.

It is already known [39] that the representations that admit wavelets are faithful and weakly equivalent to the right regular representation of the group.

Theorem 4.6.

(i) *Let H be a Hilbert space, S an isometry on H. Then there exist a Hilbert space $\hat{H}$ containing H and a unitary $\hat{S}$ on $\hat{H}$ such that*

$$\hat{S}|_H = S, \tag{4.7}$$

$$\overline{\bigcup_{n\geq 0} \hat{S}^{-n} H} = \hat{H}. \tag{4.8}$$

Moreover these are unique up to an intertwining isomorphism.

(ii) *If $\mathcal{A}$ is a C^*-algebra, α is an endomorphism on $\mathcal{A}$ and π is a representation of $\mathcal{A}$ on H such that*

$$S\pi(g) = \pi(\alpha(g))S, \qquad g \in \mathcal{A}, \tag{4.9}$$

then there exists a unique representation $\hat{\pi}$ on $\hat{H}$ such that

$$\hat{\pi}(g)|_H = \pi(g), \qquad g \in \mathcal{A}, \tag{4.10}$$

$$\hat{S}\hat{\pi}(g) = \hat{\pi}(\alpha(g))\hat{S}, \qquad g \in \mathcal{A}. \tag{4.11}$$

Corollary 4.7. *Let X, r, and μ be as in Corollary 4.5. Let I be a finite or countable set. Suppose $H \colon X \to \mathcal{B}(\ell^2(I))$ has the property that $H(x) \geq 0$ for almost every $x \in X$, and $H_{ij} \in L^1(X)$ for all $i, j \in I$. Let $M_0 \colon X \to \mathcal{B}(\ell^2(I))$ such that $x \mapsto \|M_0(x)\|$ is essentially bounded. Assume in addition that*

$$\frac{1}{\# r^{-1}(x)} \sum_{r(y)=x} M_0^*(y) H(y) M_0(y) = H(x), \quad \text{for a.e. } x \in X. \tag{4.12}$$

Then there exists a Hilbert space $\hat{K}$, a unitary operator $\hat{U}$ on $\hat{K}$, a representation $\hat{\pi}$ of $L^\infty(X)$ on $\hat{K}$, and a family of vectors $(\varphi_i) \in \hat{K}$, such that

$$\hat{U}\hat{\pi}(g)\hat{U}^{-1} = \hat{\pi}(g \circ r), \qquad g \in L^\infty(X),$$

$$\hat{U}\varphi_i = \sum_{j\in I} \hat{\pi}((M_0)_{ji})\varphi_j, \qquad i \in I,$$

$$\langle\, \varphi_i \mid \hat{\pi}(f)\varphi_j \,\rangle = \int_X f H_{ij}\, d\mu, \qquad i, j \in I,\ f \in L^\infty(X),$$

$$\overline{\text{span}}\{\, \hat{\pi}(f)\varphi_i \mid n \geq 0, f \in L^\infty(X), i \in I \,\} = \hat{K}.$$

These are unique up to an intertwining unitary isomorphism. (All functions are assumed weakly measurable in the sense that $x \mapsto \langle\, \xi \mid F(x)\eta \,\rangle$ is measurable for all $\xi, \eta \in \ell^2(I)$.)

Suppose now that H is a Hilbert space with an isometry S on it and with a normal representation π of $L^\infty(X)$ on H that satisfies the covariance relation

$$S\pi(g) = \pi(g \circ r)S, \qquad g \in L^\infty(X). \tag{4.13}$$

Theorem 4.6 shows that there exists a Hilbert space $\hat{H}$ containing H, a unitary $\hat{S}$ on $\hat{H}$ and a representation $\hat{\pi}$ of $L^\infty(X)$ on $\hat{H}$ such that

$(V_n := \hat{S}^{-n}(H))_n$ form an increasing sequence of subspaces with dense union,

$$\hat{S}|_H = S,$$
$$\hat{\pi}|_H = \pi,$$
$$\hat{S}\hat{\pi}(g) = \hat{\pi}(g \circ r)\hat{S}.$$

Theorem 4.8.

(i) $V_1 = \hat{S}^{-1}(H)$ *is invariant for the representation* $\hat{\pi}$*. The multiplicity functions of the representation* $\hat{\pi}$ *on* V_1*, and on* $V_0 = H$*, are related by*

$$m_{V_1}(x) = \sum_{r(y)=x} m_{V_0}(y), \qquad x \in X. \tag{4.14}$$

(ii) *If* $W_0 := V_1 \ominus V_0 = \hat{S}^{-1}H \ominus H$*, then*

$$m_{V_0}(x) + m_{W_0}(x) = \sum_{r(y)=x} m_{V_0}(y), \qquad x \in X. \tag{4.15}$$

Proof. Note that $\hat{S}$ maps V_1 to V_0, and the covariance relation implies that the representation $\hat{\pi}$ on V_1 is isomorphic to the representation $\pi^r : g \mapsto \pi(g \circ r)$ on V_0. Therefore we have to compute the multiplicity of the latter, which we denote by $m^r_{V_0}$.

By the spectral theorem there exists a unitary isomorphism $J\colon H(= V_0) \to L^2(X, m_{V_0}, \mu)$, where, for a *multiplicity function* $m\colon X \to \{0, 1, \dots, \infty\}$, we use the notation

$$L^2(X, m, \mu) := \{\, f\colon X \to \cup_{x\in X}\mathbb{C}^{m(x)} \mid f(x) \in \mathbb{C}^{m(x)}, \int_X \|f(x)\|^2 \, d\mu(x) < \infty \,\}.$$

In addition J intertwines π with the representation of $L^\infty(X)$ by multiplication operators, i.e.,

$$(J\pi(g)J^{-1}(f))(x) = g(x)f(x), \qquad g \in L^\infty(X), f \in L^2(X, m_{V_0}, \mu), x \in X.$$

Remark 4.9. Here we are identifying H with $L^2(X, m_{V_0}, \mu)$ via the *spectral representation*. We recall the details of this representation $H \ni f \mapsto \tilde{f} \in L^2(X, m_{V_0}, \mu)$.

Recall that any normal representation $\pi \in \mathrm{Rep}(L^\infty(X), H)$ is the orthogonal sum

$$H = \sum_{k\in C}^{\oplus} [\pi(L^\infty(X))k], \tag{4.16}$$

where the set C of vectors $k \in H$ is chosen such that

- $$\|k\| = 1,$$
- $$\langle\, k \mid \pi(g)k \,\rangle = \int_X g(x) v_k(x)^2 \, d\mu(x), \text{ for all } k \in C;$$
- $$\langle\, k' \mid \pi(g)k \,\rangle = 0, \quad g \in L^\infty(X), k, k' \in C, k \neq k'; \text{ orthogonality.}$$

The formula (4.16) is obtained by a use of Zorn's lemma. Here, v_k^2 is the Radon-Nikodym derivative of $\langle\, k \mid \pi(\cdot)k \,\rangle$ with respect to μ, and we use that π is assumed normal.

For $f \in H$, set

$$f = \sum_{k\in C}^{\oplus} \pi(g_k)k, \quad g_k \in L^\infty(X)$$

and

$$\tilde{f} = \sum_{k\in C}^{\oplus} g_k v_k \in L^2_\mu(X, \ell^2(C)).$$

Then $Wf = \tilde{f}$ is the desired spectral transform, i.e.,

$$W \text{ is unitary,}$$

$$W\pi(g) = M(g)W,$$

and

$$\|\tilde{f}(x)\|^2 = \sum_{k\in C} |g_k(x)v_k(x)|^2.$$

Indeed, we have

$$\int_X \|\tilde{f}(x)\|^2 \, d\mu(x) = \int_X \sum_{k\in C} |g_k(x)|^2 v_k(x)^2 \, d\mu(x) = \sum_{k\in C} \int_X |g_k|^2 v_k^2 \, d\mu$$

$$= \sum_{k\in C} \langle\, k \mid \pi(|g_k|^2)k \,\rangle = \sum_{k\in C} \|\pi(g_k)k\|^2 = \left\| \sum_{k\in C}^{\oplus} \pi(g_k)k \right\|_H^2 = \|f\|_H^2.$$

It follows in particular that the multiplicity function $m(x) = m_H(x)$ is

$$m(x) = \#\{\, k \in C \mid v_k(x) \neq 0 \,\}.$$

Setting

$$X_i := \{\, x \in X \mid m(x) \geq i \,\}, \qquad i \geq 1,$$

we see that

$$H \simeq \sum^{\oplus} L^2(X_i, \mu) \simeq L^2(X, m, \mu),$$

and the isomorphism intertwines $\pi(g)$ with multiplication operators.

Returning to the proof of the theorem, we have to find the similar form for the representation π^r. Let

$$\tilde{m}(x) := \sum_{r(y)=x} m_{V_0}(y), \qquad x \in X. \tag{4.17}$$

Define the following unitary isomorphism:

$$L : L^2(X, m_{V_0}, \mu) \to L^2(X, \tilde{m}, \mu),$$

$$(L\xi)(x) = \frac{1}{\sqrt{\# r^{-1}(x)}}(\xi(y))_{r(y)=x}.$$

(Note that the dimensions of the vectors match because of (4.17).) This operator L is unitary. For $\xi \in L^2(X, m_{V_0}, \mu)$, we have

$$\begin{aligned}\|L\xi\|^2_{L^2(X,m_{V_0},\mu)} &= \int_X \|L\xi(x)\|^2 \, d\mu(x) \\ &= \int_X \frac{1}{\# r^{-1}(x)} \sum_{r(y)=x} \|\xi(y)\|^2 \, d\mu(x) \\ &= \int_X \|\xi(x)\|^2 \, d\mu(x).\end{aligned}$$

And L intertwines the representations. Indeed, for $g \in L^\infty(X)$,

$$L(g \circ r \, \xi)(x) = (g(r(y))\xi(y))_{r(y)=x} = g(x)L(\xi)(x).$$

Therefore, the multiplicity of the representation $\pi^r : g \mapsto \pi(g \circ r)$ on V_0 is $\tilde{m}$, and this proves (i).

(ii) follows from (i).

Conclusions. By definition, if $k \in C$,

$$\langle\, k \mid \pi(g)k \,\rangle = \int_X g(x) v_k(x)^2 \, d\mu(x), \text{ and}$$

$$\langle\, k \mid \pi^r(g)k \,\rangle = \int_X g(r(x)) v_k(x)^2 \, d\mu(x) = \int_X g(x) \frac{1}{\# r^{-1}(x)} \sum_{r(y)=x} v_k(x)^2 \, d\mu(x);$$

and so

$$\begin{aligned} m^r(x) &= \# \{\, k \in C \mid \sum_{r(y)=x} v_k(y)^2 > 0 \,\} \\ &= \sum_{r(y)=x} \# \{\, k \in C \mid v_k(y)^2 > 0 \,\} = \sum_{r(y)=x} m(y). \end{aligned}$$

Let $C^m(x) := \{\, k \in C \mid v_k(x) \neq 0 \,\}$. Then we showed that

$$C^m(x) = \bigcup_{y \in X, r(y)=x} C^m(y)$$

and that $C^m(y) \cap C^m(y') = \emptyset$ when $y \neq y'$ and $r(y) = r(y') = x$. Setting $\mathcal{H}(x) = \ell^2(C^m(x))$, we have

$$\mathcal{H}(x) = \ell^2(C^m(x)) = \sum_{r(y)=x}^{\oplus} \ell^2(C^m(y)) = \sum_{r(y)=x}^{\oplus} \mathcal{H}(y). \qquad \square$$

4.1. Spectral decomposition of covariant representations: projective limits

We give now a different representation of the construction of the covariant system associated to m_0 and h.

We work in either the category of measure spaces or topological spaces.

Definition 4.10. Let $r\colon X \to X$ be onto, and assume that $\# r^{-1}(x) < \infty$ for all $x \in X$. We define the *projective limit* of the system:

$$X \xleftarrow{r} X \xleftarrow{r} X \cdots \to X_\infty \,, \tag{4.18}$$

as

$$X_\infty := \{\, \hat{x} = (x_0, x_1, \dots) \mid r(x_{n+1}) = x_n, \text{ for all } n \geq 0 \,\}.$$

Let $\theta_n\colon X_\infty \to X$ be the projection onto the nth component:

$$\theta_n(x_0, x_1, \dots) = x_n, \qquad (x_0, x_1, \dots) \in X_\infty.$$

Taking inverse images of sets in X through these projections, we obtain a sigma-algebra on X_∞, or a topology on X_∞.

We have an induced mapping $\hat{r}\colon X_\infty \to X_\infty$ defined by

$$\hat{r}(\hat{x}) = (r(x_0), x_0, x_1, \dots), \text{ and with inverse } \hat{r}^{-1}(\hat{x}) = (x_1, x_2, \dots). \tag{4.19}$$

so $\hat{r}$ is an automorphism, i.e., $\hat{r} \circ \hat{r}^{-1} = \mathrm{id}_{X_\infty}$ and $\hat{r}^{-1} \circ \hat{r} = \mathrm{id}_{X_\infty}$.

Note that

$$\theta_n \circ \hat{r} = r \circ \theta_n = \theta_{n-1},$$

Consider a probability measure μ on X that satisfies

$$\int_X f \, d\mu = \int_X \frac{1}{\# r^{-1}(x)} \sum_{r(y)=x} f(y) \, d\mu(x). \tag{4.20}$$

For $m_0 \in L^\infty(X)$, define

$$(R\xi)(x) = \frac{1}{\# r^{-1}(x)} \sum_{r(y)=x} |m_0(y)|^2 \xi(y), \qquad \xi \in L^1(X). \tag{4.21}$$

We now resume the operator theoretic theme of our paper, that of extending a system of operators in a given Hilbert space to a "dilated" system in an ambient or extended Hilbert space; the general idea being that the dilated system acquires a "better" spectral theory; for example, contractive operators dilate to unitary operators in the extended Hilbert space; and similarly, endomorphisms dilate to automorphisms. This of course is a general theme in operator theory; see e.g., [10] and [38]. But in our present setting, there is much more structure than the mere Hilbert space geometry: We must adapt the underlying operator theory to the particular function spaces and measures at hand. The next theorem (Theorem 4.13) is key to the understanding of the dilation step as it arises in our context of multiscale theory. It takes into consideration how we arrive at our function spaces by making dynamics out of the multiscale setting. Hence, the dilations we construct take place at three separate levels as follows:

- dynamical systems,

$$(X, r, \mu) \text{ endomorphism } \to (X_\infty, \hat{r}, \hat{\mu}), \text{ automorphism };$$

- Hilbert spaces,

$$L_2(X, h\, d\mu) \to (R_{m_0}h = h) \to L^2(X_\infty, \hat{\mu});$$

- operators,

$$S_{m_0} \text{ isometry } \to U \text{ unitary (if } m_0 \text{ is non-singular)},$$

$$M(g) \text{ multiplication operator } \to M_\infty(g).$$

Definition 4.11. A function m_0 on a measure space is called *singular* if $m_0 = 0$ on a set of positive measure.

In general, the operators S_{m_0} on $H_0 = L^2(X, h\, d\mu)$, and U on $L^2(X_\infty, \hat{\mu})$, may be given only by abstract Hilbert-space axioms; but in our *martingale representation*, we get the following two concrete formulas:

$$(S_{m_0}\xi)(x) = m_0(x)\xi(r(x)), \qquad x \in X, \xi \in H_0,$$

$$(Uf)(\hat{x}) = m_0(x_0)f(\hat{r}(\hat{x})), \qquad \hat{x} \in X_\infty, f \in L^2(X_\infty, \hat{\mu}).$$

Definition 4.12. Let X, r, μ, and W be given as before. Recall $r\colon X \to X$ is a finite-to-one map onto X, μ is a corresponding strongly invariant measure, and $W\colon X \to [0, \infty)$ is a measurable weight function. We define the corresponding Ruelle operator $R = R_W$, and we let h be a chosen Perron–Frobenius–Ruelle eigenfunction. From these data, we define the following sequence of measures ω_n on X. (Our presentation of a wavelet transform in Theorem 4.15 will depend on this sequence.) For each n, the measure ω_n is defined by the following formula:

$$\omega_n(f) = \int_X R^n(fh)\, d\mu, \qquad f \in L^\infty(X). \tag{4.22}$$

To build our wavelet transform, we first prove in Theorem 4.13 below that each of the measure families (ω_n) defines a unique W-quasi-invariant measure $\hat{\mu}$ on X_∞. In the theorem, we take $W := |m_0|^2$, and we show that $\hat{\mu}$ is quasi-invariant with respect to $\hat{r}$ with transformation Radon-Nikodym derivative W.

Theorem 4.13. *If $h \in L^1(X)$, $h \geq 0$ and $Rh = h$, then there exists a unique measure $\hat{\mu}$ on X_∞ such that*

$$\hat{\mu} \circ \theta_n^{-1} = \omega_n, \qquad n \geq 0.$$

We can identify functions on X with functions on X_∞ by

$$f(x_0, x_1, \dots) = f(x_0), \qquad f\colon X \to \mathbb{C}.$$

Under this identification,

$$\frac{d(\hat{\mu} \circ \hat{r})}{d\hat{\mu}} = |m_0|^2. \tag{4.23}$$

Theorem 4.14. *Suppose m_0 is non-singular, i.e., it does not vanish on a set of positive measure. Define U on $L^2(X_\infty, \hat{\mu})$ by*

$$
\begin{aligned}
Uf &= m_0 f \circ \hat{r}, \qquad f \in L^2(X_\infty, \hat{\mu}),\\
\pi(g)f &= gf, \qquad g \in L^\infty(X), f \in L^2(X_\infty, \hat{\mu}),\\
\varphi &= 1.
\end{aligned}
$$

Then $(L^2(X_\infty, \hat{\mu}), U, \pi, \varphi)$ is the covariant system associated to m_0 and h as in Corollary 4.5. Moreover, if $M_g f = gf$ for $g \in L^\infty(X_\infty, \hat{\mu})$ and $f \in L^2(X_\infty, \hat{\mu})$, then

$$UM_gU^{-1} = M_{g\circ\hat{r}}.$$

The Hilbert space $L^2(X_\infty, \hat{\mu})$ admits a different representation as an L^2-martingale Hilbert space. Let

$$H_n := \{\, f \in L^2(X_\infty, \hat{\mu}) \mid f = \xi \circ \theta_n, \xi \in L^2(X, \omega_n) \,\}.$$

Then H_n form an increasing sequence of closed subspaces which have dense union.

We can identify the functions in H_n with functions in $L^2(X, \omega_n)$, by

$$i_n(\xi) = \xi \circ \theta_n, \qquad \xi \in L^2(X, \omega_n).$$

The definition of $\hat{\mu}$ makes i_n an isomorphism between H_n and $L^2(X, \omega_n)$.

Define

$$\mathcal{H} := \left\{ (\xi_0, \xi_1, \dots) \;\middle|\; \xi_n \in L^2(X, \omega_n), R(\xi_{n+1}h) = \xi_n h,\ \sup_n \int_X R^n(|\xi_n|^2 h)\, d\mu < \infty \right\}$$

with the scalar product

$$\langle\, (\xi_0, \xi_1, \dots) \mid (\eta_0, \eta_1, \dots) \,\rangle = \lim_{n\to\infty} \int_X R^n(\overline{\xi}_n \eta_n h)\, d\mu.$$

Theorem 4.15. *The map $\Phi\colon L^2(X_\infty, \hat{\mu}) \to \mathcal{H}$ defined by*

$$\Phi(f) = (i_n^{-1}(P_n f))_{n\geq 0},$$

where P_n is the projection onto H_n, is an isomorphism. The transform Φ satisfies the following three conditions, and it is determined uniquely by them:

$$
\begin{aligned}
\Phi U \Phi^{-1}(\xi_n)_{n\geq 0} &= (m_0 \circ r^n\, \xi_{n+1})_{n\geq 0},\\
\Phi \pi(g) \Phi^{-1}(\xi_n)_{n\geq 0} &= (g \circ r^n\, \xi_n)_{n\geq 0},\\
\Phi\varphi &= (1, 1, \dots).
\end{aligned}
$$

Theorem 4.16. *There exists a unique isometry $\Psi\colon L^2(X_\infty, \hat{\mu}) \to \tilde{H}$ such that*

$$\Psi(\xi \circ \theta_n) = \tilde{U}^{-n} \tilde{\pi}(\xi) \tilde{U}^n \tilde{\varphi}, \qquad \xi \in L^\infty(X, \mu).$$

Ψ intertwines the two systems, i.e.,

$$\Psi U = \tilde{U}\Psi,\ \Psi\pi(g) = \tilde{\pi}(g)\Psi, \quad \text{for } g \in L^\infty(X, \mu),\ \Psi\varphi = \tilde{\varphi}.$$

Theorem 4.17. *Let (m_0, h) be a Perron–Ruelle–Frobenius pair with m_0 non-singular.*

(i) *For each operator A on $L^2(X_\infty, \hat{\mu})$ which commutes with U and π, there exists a cocycle f, i.e., a bounded measurable function $f\colon X_\infty \to \mathbb{C}$ with $f = f \circ \hat{r}$, $\hat{\mu}$-a.e., such that*
$$A = M_f, \tag{4.24}$$
and, conversely each cocycle defines an operator in the commutant.

(ii) *For each measurable harmonic function $h_0\colon X \to \mathbb{C}$, i.e., $R_{m_0} h_0 = h_0$, with $|h_0|^2 \leq ch^2$ for some $c \geq 0$, there exists a unique cocycle f such that*
$$h_0 = E_0(f)h, \tag{4.25}$$
where E_0 denotes the conditional expectation from $L^\infty(X_\infty, \hat{\mu})$ to $L^\infty(X, \mu)$, and conversely, for each cocycle the function h_0 defined by (4.25) *is harmonic.*

(iii) *The correspondence $h_0 \to f$ in* (ii) *is given by*
$$f = \lim_{n\to\infty} \frac{h_0}{h} \circ \theta_n \tag{4.26}$$
where the limit is pointwise $\hat{\mu}$-a.e., and in $L^p(X_\infty, \hat{\mu})$ for all $1 \leq p < \infty$.

The final result in this section concerns certain path-space measures, indexed by a base space X. Such a family is also called a process: it is a family of positive Radon measures P_x, indexed by x in the base space X. Each P_x is a measure on the probability space Ω which is constructed from X by the usual Cartesian product, i.e., countably infinite Cartesian product of X with itself. The Borel structure on Ω is generated by the usual cylinder subsets of Ω. Given a weight function W on X, the existence of the measures P_x comes from an application of a general principle of Kolmogorov.

Let X be a metric space and $r\colon X \to X$ an N to 1 map. Denote by $\tau_k\colon X \to X$, $k \in \{1, \dots, N\}$, the branches of r, i.e., $r(\tau_k(x)) = x$ for $x \in X$, the sets $\tau_k(X)$ are disjoint and they cover X.

Let μ be a measure on X with the property
$$\mu = \frac{1}{N} \sum_{k=1}^{N} \mu \circ \tau_k^{-1}. \tag{4.27}$$
This can be rewritten as
$$\int_X f(x)\, d\mu(x) = \frac{1}{N} \sum_{k=1}^{N} \int_X f(\tau_k(x))\, d\mu(x), \tag{4.28}$$
which is equivalent also to the strong invariance property.

Let W, $h \geq 0$ be two functions on X such that
$$\sum_{k=1}^{N} W(\tau_k(x)) h(\tau_k(x)) = h(x), \qquad x \in X. \tag{4.29}$$
Denote by Ω the multi-index set
$$\Omega := \Omega_N := \prod_{\mathbb{N}} \{1, \dots, N\}.$$

Also we define

$$W^{(n)}(x) := W(x)W(r(x)) \cdots W(r^{n-1}(x)), \qquad x \in X.$$

The final result in this section concerns certain path-space measures. In this context each measure P_x is a measure on Ω. We will return to a more general variant of these path-space measures in Theorem 5.12 in the next section. This will be in the context of a system (X, r, V) where $r\colon X \to X$ is a give finite to one endomorphism, and $W\colon X \to [0, \infty)$ is a measurable weight function. In this context the measures (P_x) are indexed by a base space X, and they are measures on the projective space X_∞ built on (X, r), see Definition 4.10. The family (P_x) is also called a process: each P_x is a positive Radon measure on X_∞, and X_∞ may be thought of as a discrete path space.

The Borel structure on Ω, and on X_∞, is generated by the usual cylinder subsets. Given a weight function W on X, the existence of the measures $P_x := P_x^W$ comes from an application of a general principle of Kolmogorov.

Proposition 4.18. *For every $x \in X$ there exists a positive Radon measure P_x on Ω such that, if f is a bounded measurable function on X which depends only on the first n coordinates $\omega_1, \dots, \omega_n$, then*

$$\int_\Omega f(\omega)\, dP_x(\omega)$$
$$= \sum_{\omega_1,\dots,\omega_n} W^{(n)}(\tau_{\omega_n}\tau_{\omega_{n-1}} \cdots \tau_{\omega_1}(x))h(\tau_{\omega_n}\tau_{\omega_{n-1}} \cdots \tau_{\omega_1}(x))f(\omega_1, \dots, \omega_n). \quad (4.30)$$

5. Remarks on other applications

Wavelet sets. Investigate the existence and construction of wavelet sets and elementary wavelets and frames in the Julia-set theory, and the corresponding interpolation theory, and their relationships to generalized multiresolution analysis (GMRA). The unitary system approach to wavelets by Dai and Larson [32] dealt with systems that can be very irregular. And recent work by Ólafsson et al. shows that wavelet sets can be essential to a basic dilation-translation wavelet theory even for a system where the set of dilation unitaries is not a group.

The renormalization question. When are there renormalizable iterates of r in arithmetic progression, i.e., when are there iteration periods n such that the system $\{r^{kn}, k \in \mathbb{N}\}$ may be *renormalized*, or rescaled to yield a new dynamical system of the same general shape as that of the original map $r(z)$?

Since the scaling equation (2.2) from wavelet theory is a renormalization, our application of multiresolutions to the Julia sets $X(r)$ of complex dynamics suggests a useful approach to renormalization in this context. The drawback of this approach is that it relies on a rather unwieldy Hilbert space built on $X(r)$, or rather on the projective system $X(r)_\infty$ built in turn over $X(r)$. So we are left with translating our Hilbert-space theoretic normalization back to the direct and geometric algorithms on the complex plane.

General and rigorous results from complex dynamics state that under additional geometric hypotheses, renormalized dynamical systems range in a compact family.

The use of geometric tools from Hilbert space seems promising for renormalization questions (see, e.g., [64], [15], [23], and [74]) since notions of repetition up to similarity of form at infinitely many scales are common in the Hilbert-space approach to multiresolutions; see, e.g., [35]. And this self-similarity up to scale parallels a basic feature of renormalization questions in physics: for a variety of instances of dynamics in physics, and in telecommunication [64, 15, 20], we encounter scaling laws of self-similarity; i.e., we observe that a phenomenon reproduces itself on different time and/or space scales.

Self-similar processes are stochastic processes that are invariant in distribution under suitable scaling of time and/or space (details below!) Fractional Brownian motion is perhaps the best known of these, and it is used in telecommunication and in stochastic integration. While the underlying idea behind this can be traced back to Kolmogorov, it is only recently, with the advent of wavelet methods, that its computational power has come more into focus, see, e.g., [64]. But at the same time, this connection to wavelet analysis is now bringing the *operator-theoretic features* in the subject to the fore.

In statistics, we observe that fractional Brownian motion (a Gaussian process $B(t)$ with $E\{B(t)\} = 0$, and covariance $E\{B(t)B(s)\}$ given by a certain h-fractional law) has the property that there is a number h such that, for all a, the two processes $B(at)$ and $a^h B(t)$ have the same finite-dimensional distributions. The scaling feature of fractional Brownian motion, and of its corresponding white-noise process, is used in telecommunication, and in stochastic integration; and this connection has been one of the new and more exciting domains of applications of wavelet tools, both pure and applied (Donoho, Daubechies, Meyer, etc. [35, 33, 63]).

This is a new direction of pure mathematics which makes essential contact with problems that are not normally investigated in the context of harmonic analysis.

In our proofs, we take advantage of our Hilbert-space formulation of the notion of *similarity up to scale*, and the use of scales of closed subspaces $\mathcal{V}_n$ in a Hilbert space $\mathcal{H}$. The unitary operator U which scales between the spaces may arise from a substitution in nonlinear dynamics, such as in the context of Julia sets [11]; it may "scale" between the sigma-algebras in a martingale [36, 37, 75]; it may be a scaling of large volumes of data (see, e.g., [1]); it may be the scaling operation in a fractional Brownian motion [64, 59]; it may be cell averages in finite elements [29]; or it may be the traditional dyadic scaling operator in $\mathcal{H} = L^2(\mathbb{R})$ in wavelet frame analysis [28, 29].

Once the system $(\mathcal{H}, (\mathcal{V}_n), U)$ is given, then it follows that there is a unitary representation V of $\mathbb{T}$ in $\mathcal{H}$ which defines a grading of operators on $\mathcal{H}$. Moreover

U then defines a similarity up to scale precisely when U has grade one, see (3.4) for definition. We also refer to the papers [4, 10] for details and applications of this notion in operator theory.

The general idea of assigning degrees to operators in Hilbert space has served as a powerful tool in other areas of mathematical physics (see, e.g., [3], [22]; and [16] on phase transition problems); operator theory [5, 4, 38, 2]; and operator algebras; see, e.g., [31, 73, 24].

Our research, e.g., [43], in fact already indicates that the nonlinear problems sketched above can be attacked with the use of our operator-theoretic framework.

Wavelet representations of operator-algebraic crossed products. The construction of wavelets requires a choice of a low-pass filter $m_0 \in L^\infty(\mathbb{T})$ and of $N-1$ high-pass filters $m_i \in L^\infty(\mathbb{T})$ that satisfy certain orthogonality conditions. As shown in [17], these choices can be put in one-to-one correspondence with a class of covariant representations of the Cuntz algebra $\mathcal{O}_N$. In the case of an arbitrary N-to-one dynamical system $r\colon X \to X$, the wavelet construction will again involve a careful choice of the filters, and a new class of representations is obtain. The question here is to classify these representations and see how they depend on the dynamical system r. A canonical choice of the filters could provide an invariant for the dynamical system.

In a recent paper [47], the crossed-product of a C^*-algebra by an endomorphism is constructed using the transfer operator. The required covariant relations are satisfied by the operators that we introduced in [43], [44].

There are two spaces for these representations: one is the core space of the multiresolution $V_0 = L^2(X, \mu)$. Here we have the abelian algebra of multiplication operators and the isometry $S\colon f \mapsto m_0\, f \circ r$. Together they provide representations of the crossed product by an endomorphism. The second space is associated to a dilated measure $\hat{\mu}$ on the solenoid of r, $L^2(X_\infty, \hat{\mu})$ (see [43], [44]). The isometry S is dilated to a unitary $\hat{S}$ which preserves the covariance. In this case, we are dealing with representations of crossed products by automorphisms.

Thus we have four objects interacting with one another: crossed products by endomorphisms and their representations, and crossed products by automorphisms and its representations on $L^2(\hat{X}, \hat{\mu})$.

The representations come with a rich structure given by the multiresolution, scaling function and wavelets. Therefore their analysis and classification seems promising.

KMS-states and equilibrium measures. In [48], the KMS states on the Cuntz–Pimsner algebra for a certain one-parameter group of automorphisms are obtained by composing the unique equilibrium measure on an abelian subalgebra (i.e., the measure which is invariant to the transfer operator) with the conditional expectation. These results were further generalized in [58] for expansive maps and the associated groupoid algebras.

As we have seen in [45] for the case when $r(z) = z^N$, the covariant representations associated to some low-pass filter m_0 are highly dependent on the equilibrium measure ν. As outlined above, we consider here an expansive dynamical system built on a given finite to one mapping $r\colon X \to X$, and generalizing the familiar case of the winding mapping $z \to z^N$ on $\mathbb{T}$. In our $r\colon X \to X$ context, we then study weight functions $V\colon X \to [0,\infty)$. If the zero-set of V is now assumed finite, we have so far generalized what is known in the $z \to z^N$ case. Our results take the form of a certain dichotomy for the admissible Perron–Frobenius–Ruelle measures, and they fit within the framework of [58]. Hence our construction also yields covariant representations and symmetry in the context of [58]. Our equilibrium measures in turn induce the kind of KMS-states studied in [58].

The results of [58] and [48] restrict the weight function (in our case represented by the absolute square of the low-pass filter, $|m_0|^2$) to being strictly positive and Hölder continuous. These restrictions are required because of the form of the Ruelle–Perron–Frobenius theorem used, which guarantees the existence and uniqueness of the equilibrium measure.

However, the classical wavelet theory shows that many interesting examples require a "low-pass condition", $m_0(1) = \sqrt{N}$, which brings zeroes for $|m_0|^2$.

Our martingale approach is much less restrictive than the conditions of [58]: it allows zeroes and discontinuities. Coupled with a recent, more general form of Ruelle's theorem from [49], we hope to extend the results of [58] and be able to find the KMS states in a more general case. Since the existence of zeroes can imply a multiplicity of equilibrium measures (see [41], [19]) a new phenomenon might occur such as spontaneously breaking symmetry.

For each KMS state, one can construct the GNS representation. Interesting results are known [65] for the subshifts of finite type and the Cuntz–Krieger algebra, when the construction yields type III_λ AFD factors.

The natural question is what type of factors appear in the general case of $r\colon X \to X$. Again, a canonical choice for the low-pass filter can provide invariants for the dynamical system.

Historically, von-Neumann-algebra techniques have served as powerful tools for analyzing discrete structures. Our work so far suggests that the iteration systems (X, r) will indeed induce interesting von-Neumann-algebra isomorphism classes.

Dimension groups. We propose to use our geometric algorithms [20, 21] for deciding order isomorphism of dimension groups for the purpose of deciding isomorphism questions which arise in our multiresolution analysis built on Julia sets. The project is to use the results in [20] to classify substitution tilings in the sense of [66], [67].

Equilibrium measures, harmonic functions for the transfer operator, and infinite Riesz products. In [19], [41], for the dynamical system $z \mapsto z^N$, we were able to compute the equilibrium measures, the fixed points of the transfer operator and its ergodic decomposition by analyzing the commutant of the covariant representations associated to the filter m_0. In [43], [44] we have shown that also in the general case there is a one-to-one correspondence between harmonic functions for the transfer operator and the commutant. Thus an explicit form of the covariant representation can give explicit solutions for the eigenvalue problem $R_{m_0}h = h$, which are of interest in ergodic theory (see [9]).

In some cases, the equilibrium measures are Riesz products (see [45]); these are examples of exotic measures which arise in harmonic analysis [12, 46, 56, 6, 69, 14, 26, 27, 57, 62, 70]. The wavelet-operator-theoretic approach may provide new insight into their properties.

Non-uniqueness of the Ruelle–Perron–Frobenius data. A substantial part of the current literature in the Ruelle–Perron–Frobenius operator in its many guises is primarily about conditions for uniqueness of KMS; so that means no phase transition. This is ironic since Ruelle's pioneering work was motivated by phase-transition questions, i.e., non-uniqueness.

In fact non-uniqueness is much more interesting in a variety of applications: That is when we must study weight functions W in the Ruelle operator $R = R_W$ when the standard rather restrictive conditions on W are in force. The much celebrated Ruelle theorem for the transition operator R gives existence and uniqueness of Perron–Frobenius data under these restrictive conditions. But both demands from physics, and from wavelets, suggest strong and promising research interest in relaxing the stringent restrictions that are typical in the literature, referring to assumptions on W. There is a variety of exciting possibilities. They would give us existence, but not necessarily uniqueness in the conclusion of a new Ruelle-type theorem.

Induced measures. A basic tool in stochastic processes (from probability theory) involves a construction on a "large" projective space X_∞, based on some marginal measure on some coordinate space X. Our projective limit space X_∞ will be constructed from a finite branching process.

Let A be a $k \times k$ matrix with entries in $\{0, 1\}$. Suppose every column in A contains an entry 1.

Set $X(A) := \left\{ (\xi_i)_{i\in\mathbb{N}} \in \prod_{\mathbb{N}}\{1, \dots, k\} \;\middle|\; A(\xi_i, \xi_{i+1}) = 1 \right\}$ and

$$r_A(\xi_1, \xi_2, \dots) = (\xi_2, \xi_3, \dots) \quad \text{for } \xi \in X(A).$$

Then r_A is a subshift, and the pair $(X(A), r_A)$ satisfies our conditions.

It is known [72] that, for each A, as described, the corresponding system $r_A \colon X(A) \to X(A)$ has a unique strongly r_A-invariant probability measure, ρ_A, i.e., a probability measure on $X(A)$ such that $\rho_A \circ R_1 = \rho_A$, where R_1 is defined as in (5.1) below.

We now turn to the connection between measures on X and an associated family of induced measures on X_∞, and we characterize those measures X_∞ which are quasi-invariant with respect to the invertible mapping $\hat{r}$, where $X_\infty := \{\, \hat{x} = (x_0, x_1, \dots) \in \prod_{\mathbb{N}_0} X \mid r(x_{n+1}) = x_n \,\}$, $\hat{r}(\hat{x}) = (r(x_0), x_0, x_1, \dots)$.

In our extension of measures from X to X_∞, we must keep track of the transfer from one step to the next, and there is an operator which accomplishes this, Ruelle's transfer operator

$$R_W f(x) = \frac{1}{\# r^{-1}(x)} \sum_{r(y)=x} W(y) f(y), \qquad f \in L^1(X,\mu). \tag{5.1}$$

In its original form it was introduced in [72], but since that, it has found a variety of applications, see e.g., [9]. For use of the Ruelle operator in wavelet theory, we refer to [54] and [41].

The Hilbert spaces of functions on X_∞ will be realized as a Hilbert spaces of martingales. This is consistent with our treatment of wavelet resolutions as martingales. This was first suggested in [51] in connection with wavelet analysis.

In [43], we studied the following restrictive setup: we assumed that X carries a probability measure μ which is *strongly r-invariant*. By this we mean that

$$\int_X f\,d\mu = \int_X \frac{1}{\# r^{-1}(x)} \sum_{y\in X, r(y)=x} f(y)\,d\mu(x), \qquad f \in L^\infty(X). \tag{5.2}$$

If, for example $X = \mathbb{R}/\mathbb{Z}$, and $r(x) = 2x \bmod 1$, then the Haar measure on $\mathbb{R}/\mathbb{Z}$ = Lebesgue measure on $[0,1)$ is the unique strongly r-invariant measure on X.

Suppose the weight function V is bounded and measurable. Then define $R = R_V$, the *Ruelle operator*, by formula (5.1), with $V = W$.

Theorem 5.1. ([43]) *Let $r\colon X \to X$ and $X_\infty(r)$ be as described above, and suppose that X has a strongly r-invariant measure μ. Let V be a non-negative, measurable function on X, and let R_V be the corresponding Ruelle operator.*

(i) *There is a unique measure $\hat\mu$ on $X_\infty(r)$ such that*
  (a) $\hat\mu \circ \theta_0^{-1} \ll \mu$ *(set* $h = \frac{d(\hat\mu\circ\theta_0^{-1})}{d\mu}$*),*
  (b) $\int_X f\,d\hat\mu \circ \theta_n^{-1} = \int_X R_V^n(fh)\,d\mu, \qquad n \in \mathbb{N}_0.$

(ii) *The measure $\hat\mu$ on $X_\infty(r)$ satisfies*

$$\frac{d(\hat\mu \circ \hat r)}{d\hat\mu} = V \circ \theta_0, \tag{5.3}$$

and

$$R_V h = h. \tag{5.4}$$

If the function $V\colon X \to [0,\infty)$ is given, we define

$$V^{(n)}(x) := V(x)V(r(x))\cdots V(r^{n-1}(x)),$$

and set $d\mu_n := V^{(n)} d\mu_0$. Our result states that the corresponding measure $\hat\mu$ on $X_\infty(r)$ is V-quasi-invariant if and only if

$$d\mu_0 = (V\,d\mu_0)\circ r^{-1}. \tag{5.5}$$

Theorem 5.2. *Let $V: X \to [0,\infty)$ be $\mathfrak{B}$-measurable, and let μ_0 be a measure on X satisfying the following fixed-point property*

$$d\mu_0 = (V\,d\mu_0)\circ r^{-1}. \tag{5.6}$$

Then there exists a unique measure $\hat{\mu}$ *on* $X_\infty(r)$ *such that*

$$\frac{d(\hat{\mu}\circ\hat{r})}{d\hat{\mu}} = V\circ\theta_0 \tag{5.7}$$

and

$$\hat{\mu}\circ\theta_0^{-1} = \mu_0.$$

Definition 5.3. Let $V\colon X \to [0,\infty)$ be bounded and $\mathfrak{B}$-measurable. We use the notation

$$M^V(X) := \{\,\mu \in M(X) \mid d\mu = (V\,d\mu)\circ r^{-1}\,\}.$$

For measures $\hat{\mu}$ on $(X_\infty(r), \mathfrak{B}_\infty)$ we introduce

$$M_{qi}^V(X_\infty(r)) := \left\{\,\hat{\mu}\in M(X_\infty(r)) \;\middle|\; \hat{\mu}\circ\hat{r} \ll \hat{\mu} \text{ and } \frac{d(\hat{\mu}\circ\hat{r})}{d\hat{\mu}} = V\circ\theta_0\,\right\}. \tag{5.8}$$

As in Definition 5.3, let X, r, and V be as before, i.e., r is a finite-to-one endomorphism of X, and $V\colon X\to[0,\infty)$ is a given weight function. In the next theorem, we establish a canonical bijection between the convex set $M^V(X)$ of measures on X with the set of V-quasi-invariant measures on X_∞, which we call $M_{qi}^V(X_\infty(r))$, see (5.8).

The results of the previous section may be summarized as follows:

Theorem 5.4. *Let* V *be as in Definition* 5.3. *For measures* $\hat{\mu}$ *on* $X_\infty(r)$ *and* $n\in\mathbb{N}_0$, *define*

$$C_n(\hat{\mu}) := \hat{\mu}\circ\theta_n^{-1}.$$

Then C_0 *is a bijective affine isomorphism of* $M_{qi}^V(X_\infty(r))$ *onto* $M^V(X)$ *that preserves the total measure, i.e.,* $C_0(\hat{\mu})(X) = \hat{\mu}(X_\infty(r))$ *for all* $\hat{\mu}\in M_{qi}^V(X_\infty(r))$.

Theorem 5.5. *Let* $V\colon X\to[0,\infty)$ *be continuous. Assume that there exist some measure* ν *on* $(X,\mathfrak{B})$ *and two numbers* $0<a<b$ *such that*

$$a\le\nu(X)\le b, \text{ and } a\le\int_X V^{(n)}\,d\nu\le b \text{ for all } n\in\mathbb{N}. \tag{5.9}$$

Then there exists a measure μ_0 *on* $(X,\mathfrak{B})$ *that satisfies*

$$d\mu_0 = (V\,d\mu_0)\circ r^{-1},$$

and there exists a V*-quasi-invariant measure* $\hat{\mu}$ *on* $(X_\infty(r),\mathfrak{B}_\infty)$.

Theorem 5.6. *Let* $(X,\mathfrak{B})$, *and* $r\colon X\to X$, *be as described above. Suppose* $V\colon X\to[0,\infty)$ *is measurable,*

$$\frac{1}{\#r^{-1}(x)}\sum_{r(y)=x} V(y)\le 1,$$

and that some probability measure ν_V *on* X *satisfies*

$$\nu_V\circ R_V = \nu_V. \tag{5.10}$$

Assume also that $(X,\mathfrak{B})$ *carries a strongly* r*-invariant probability measure* ρ, *such that*

$$\rho(\{\,x\in X\mid V(x)>0\,\})>0. \tag{5.11}$$

Then

(i) $T_V^n(d\rho) = R_V^n(\mathbf{1})\,d\rho$, *for* $n \in \mathbb{N}$, *where* $\mathbf{1}$ *denotes the constant function one.*
(ii) *[Monotonicity]* $\cdots \leq R_V^{n+1}(\mathbf{1}) \leq R_V^n(\mathbf{1}) \leq \cdots \leq \mathbf{1}$, *pointwise on* X.
(iii) *The limit* $\lim_{n\to} R_V^n(\mathbf{1}) = h_V$ *exists,* $R_V h_V = h_V$, *and*

$$\rho(\{\, x \in X \mid h_V(x) > 0\,\}) > 0. \tag{5.12}$$

(iv) *The measure* $d\mu_0^{(V)} = h_V\,d\rho$ *is a solution to the fixed-point problem*

$$T_V(\mu_0^{(V)}) = \mu_0^{(V)}.$$

(v) *The sequence* $d\mu_n^{(V)} = V^{(n)} h_V\,d\rho$ *defines a unique* $\hat{\mu}^{(V)}$ *as in Theorem* 5.1*; and*
(vi) $\mu_n^{(V)}(f) = \int_X R_V^n(f h_V)\,d\rho$ *for all bounded measurable functions* f *on* X, *and all* $n \in \mathbb{N}$.

Finally,

(vii) *The measure* $\hat{\mu}^{(V)}$ *on* $X_\infty(r)$ *satisfying* $\hat{\mu}^{(V)} \circ \theta_n^{-1} = \mu_n^{(V)}$ *has total mass*

$$\hat{\mu}^{(V)}(X_\infty(r)) = \rho(h_V) = \int_X h_V(x)\,d\rho(x).$$

Definition 5.7. A measure $\mu_0 \in M_1^V(X)$ is called *relatively ergodic* with respect to (r, V) if the only non-negative, bounded $\mathfrak{B}$-measurable functions f on X satisfying

$$E_{\mu_0}(Vf) = E_{\mu_0}(V) f \circ r, \ \text{ pointwise } \mu_0 \circ r^{-1}\text{-a.e.},$$

are the functions which are constant μ_0-a.e.

Since we have a canonical bijection between the two compact convex sets of measures $M_1^V(X)$ and $M_{qi,1}^V(X_\infty(r))$, the natural question arises as to the extreme points. This is answered in our next theorem. We show that there are notions ergodicity for each of the two spaces X and $X_\infty(r)$ which are equivalent to extremality in the corresponding compact convex sets of measures.

Theorem 5.8. *Let* $V\colon X \to [0, \infty)$ *be bounded and measurable. Let*

$$\hat{\mu} \in M_{qi,1}^V(X_\infty(r)), \text{ and } \mu_0 := \hat{\mu} \circ \theta_0^{-1} \in M_1^V(X).$$

The following assertions are equivalent:

(i) $\hat{\mu}$ *is an extreme point of* $M_{qi,1}^V(X_\infty(r))$*;*
(ii) $V \circ \theta_0\,d\hat{\mu}$ *is ergodic with respect to* $\hat{r}$*;*
(iii) μ_0 *is an extreme point of* $M_1^V(X)$*;*
(iv) μ_0 *is relatively ergodic with respect to* (r, V).

We now turn to the next two theorems. These are counterparts of our dimension-counting functions which we outlined above in connection with Theorem 4.8; see especially Remark 4.9. They concern the correspondence between the two classes of measures, certain on X (see Theorem 5.8), and the associated induced measures on the projective space X_∞. Our proofs are technical and will not be included. (The reader is referred to [42] for further details.) Rather we only give a few suggestive hints: Below we outline certain combinatorial concepts and functions which are central for the arguments. Since they involve a counting principle, they have an intuitive flavor which we hope will offer some insight into our theorems.

Let X be a non-empty set, $\mathfrak{B}$ a sigma-algebra of subsets of X, and $r\colon X \to X$ an onto, finite-to-one, and $\mathfrak{B}$-measurable map.

We will assume in addition that we can label measurably the branches of the inverse of r. By this, we mean that the following conditions are satisfied:

$$\text{The map} \quad \mathfrak{c}\colon X \ni x \mapsto \# r^{-1}(x) < \infty \text{ is measurable} . \tag{5.13}$$

We denote by A_i the set

$$A_i := \{ x \in X \mid \mathfrak{c}(x) = \# r^{-1}(x) \geq i \}, \qquad i \in \mathbb{N}.$$

Equation (5.13) implies that the sets A_i are measurable. Also they form a decreasing sequence and, since the map is finite-to-one,

$$X = \bigcup_{i=1}^{\infty} (A_{i+1} \setminus A_i).$$

Then, we assume that there exist measurable maps $\tau_i\colon A_i \to X$, $i \in \{1, 2, \dots\}$ such that

$$r^{-1}(x) = \{\tau_1(x), \dots, \tau_{\mathfrak{c}(x)}(x)\}, \qquad x \in X, \tag{5.14}$$

$$\tau_i(A_i) \in \mathfrak{B} \quad \text{for all } i \in \{1, 2, \dots\}. \tag{5.15}$$

Thus $\tau_1(x), \dots, \tau_{\mathfrak{c}(x)}(x)$ is a list without repetitions of the "roots" of x, $r^{-1}(x)$.

From (5.14) we obtain also that

$$\tau_i(A_i) \cap \tau_j(A_j) = \emptyset, \text{ if } i \neq j, \tag{5.16}$$

and

$$\bigcup_{i=1}^{\infty} \tau_i(A_i) = X. \tag{5.17}$$

In the sequel, we will use the following notation: for a function $f\colon X \to \mathbb{C}$, we denote by $f \circ \tau_i$ the function

$$f \circ \tau_i(x) := \begin{cases} f(\tau_i(x)) & \text{if } x \in A_i, \\ 0 & \text{if } x \in X \setminus A_i. \end{cases}$$

Our Theorem 4.13 depends on the existence of some strongly invariant measure μ on X, when the system (X, r) is given. However, in the general measurable category, such a strongly invariant measure μ on X may in fact not exist; or if it does, it may not be available by computation. In fact, recent wavelet results (for frequency localized wavelets, see, e.g., [7] and [8]) suggest the need for theorems in the more general class of dynamical systems (X, r).

In the next theorem (Theorem 5.9), we provide for each system, X, r, and V a substitute for the existence of strongly invariant measures. We show that there is a certain fixed-point property for measures on X which depends on V but not on the a priori existence of strongly r-invariant measures, and which instead is determined by a certain modular function Δ on X. This modular function in turn allows us to prove a dis-integration theorem for our V-quasi-invariant measures $\hat{\mu}$ on X_∞. In Theorem 5.12, we give a formula for this dis-integration of a V-quasi-invariant measure $\hat{\mu}$ on X_∞ in the presence of a modular function Δ. Our dis-integration of $\hat{\mu}$ is over a Markov process P_x, for x in X, which depends on Δ, but otherwise is analogous to the process P_x we used in Proposition 4.18.

Theorem 5.9. *Let $(X, \mathfrak{B})$ and $r\colon X \to X$ be as above. Let $V\colon X \to [0, \infty)$ be a bounded $\mathfrak{B}$-measurable map. For a measure μ_0 on $(X, \mathfrak{B})$, the following assertions are equivalent.*

(i) *The measure μ_0 has the fixed-point property*

$$\int_X V\, f \circ r \, d\mu_0 = \int_X f \, d\mu_0, \ \text{for all } f \in L^\infty(X, \mu_0). \tag{5.18}$$

(ii) *There exists a non-negative, $\mathfrak{B}$-measurable map Δ (depending on V and μ_0) such that*

$$\sum_{r(y)=x} \Delta(y) = 1, \ \text{for } \mu_0\text{-a.e. } x \in X, \tag{5.19}$$

and

$$\int_X V\, f \, d\mu_0 = \int_X \sum_{r(y)=x} \Delta(y) f(y) \, d\mu_0(x), \ \text{for all } f \in L^\infty(X, \mu_0). \tag{5.20}$$

Moreover, when the assertions are true, Δ is unique up to $V\, d\mu_0$-measure zero.

We recall the definitions: if $\mathfrak{B}$ is a sigma-algebra on a set X and $r\colon X \to X$ is a finite-to-one, onto and measurable map, we denote by X_∞ the set

$$X_\infty(r) := \left\{ (x_0, x_1, \dots) \in \prod_{n \in \mathbb{N}_0} X \;\middle|\; r(x_{n+1}) = x_n, \ \text{for all } n \in \mathbb{N}_0 \right\}.$$

We denote the projections by $\theta_n\colon X_\infty(r) \ni (x_0, x_1, \dots) \mapsto x_n \in X$. The union of the pull-backs $\theta_n^{-1}(\mathfrak{B})$ generates a sigma-algebra $\mathfrak{B}_\infty$. The map r extends to a measurable bijection $\hat{r}\colon X_\infty(r) \to X_\infty(r)$ defined by

$$\hat{r}(x_0, x_1, \dots) = (r(x_0), x_0, x_1, \dots).$$

Let $V\colon X \to [0, \infty)$ be a measurable, bounded function. We say that a measure μ_0 on $(X, \mathfrak{B})$ has the *fixed-point property* if

$$\int_X V\, f \circ r \, d\mu_0 = \int_X f \, d\mu_0, \qquad f \in L^\infty(X, \mu_0).$$

We say that a measure $\hat{\mu}$ on $(X_\infty(r), \mathfrak{B}_\infty)$ is V*-quasi-invariant* if

$$d(\hat{\mu} \circ \hat{r}) = V \circ \theta_0 \, d\hat{\mu}.$$

We recall the following result from [44].

Theorem 5.10. *There exists a one-to-one correspondence between measures μ_0 on X with the fixed-point property and V-quasi-invariant measures $\hat{\mu}$ on $X_\infty(r)$, given by*

$$\mu_0 = \hat{\mu} \circ \theta_0^{-1}.$$

Proposition 5.11. *Let $(X, \mathfrak{B})$, $r\colon X \to X$ and be as above, and let $D\colon X \to [0, \infty)$ be a measurable function with the property that*

$$\sum_{r(y)=x} D(y) = 1. \tag{5.21}$$

Denote by $D^{(n)}$ the product of compositions

$$D^{(n)} := D \cdot D \circ r \cdot \cdots \cdot D \circ r^{n-1}, \qquad n \in \mathbb{N}, \quad D^{(0)} := 1. \tag{5.22}$$

Then for each $x_0 \in X$, there exists a Radon probability measure P_{x_0} on Ω_{x_0} such that, if f is a bounded measurable function on Ω_{x_0} which depends only on the first n coordinates $x_1, \dots, x_n$, then

$$\int_{\Omega_{x_0}} f(\omega)\, dP_{x_0}(\omega) = \sum_{r^n(x_n)=x_0} D^{(n)}(x_n) f(x_1, \dots, x_n). \tag{5.23}$$

Theorem 5.12. *Let $(X, \mathfrak{B})$, $r\colon X \to X$ and $V\colon X \to [0, \infty)$ be as above. Let μ_0 be a measure on $(X, \mathfrak{B})$ with the fixed-point property* (5.18)*. Let $\Delta\colon X \to [0, 1]$ be the function associated to V and μ_0 as in Theorem* 5.9*, and let $\hat{\mu}$ be the unique V-quasi-invariant measure on $X_\infty(r)$ with*

$$\hat{\mu} \circ \theta_0^{-1} = \mu_0,$$

as in Theorem 5.10*. For Δ define the measures P_{x_0} as in Proposition* 5.11*. Then, for all bounded measurable functions f on $X_\infty(r)$,*

$$\int_{X_\infty(r)} f\, d\hat{\mu} = \int_X \int_{\Omega_{x_0}} f(x_0, \omega)\, dP_{x_0}(\omega)\, d\mu_0(x_0). \tag{5.24}$$

Hausdorff measure and wavelet bases. In this problem we propose wavelet bases in the context of Hausdorff measure of fractional dimension between 0 and 1. While our fractal wavelet theory has points of similarity that it shares with the standard case of Lebesgue measure on the line, there are also sharp contrasts.

It is well known that the Hilbert spaces $L^2(\mathbb{R})$ has a rich family of orthonormal bases of the following form: $\psi_{j,k}(x) = 2^{j/2}\psi(2^j x - k)$, $j, k \in \mathbb{Z}$, where ψ is a single function $\in L^2(\mathbb{R})$. Take for example Haar's function $\psi(x) = \chi_I(2x) - \chi_I(2x - 1)$ where $I = [0, 1]$ is the unit interval. Let $\mathbf{C}$ be the middle-third Cantor set. Then the corresponding indicator function $\varphi_{\mathbf{C}} := \chi_{\mathbf{C}}$ satisfies the scaling identity (see (2.2)), $\varphi_{\mathbf{C}}(\frac{x}{3}) = \varphi_{\mathbf{C}}(x) + \varphi_{\mathbf{C}}(x - 2)$.

In [45] we use this as the starting point for a multiresolution construction in a separable Hilbert space built with Hausdorff measure, and we identify the two mother functions which define the associated wavelet ONB.

Since both constructions, the first one for the Lebesgue measure, and the second one for the Hausdorff version $(dx)^s$, arise from scaling and subdivision, it seems reasonable to expect multiresolution wavelets also in Hilbert spaces constructed on the scaled Hausdorff measures $\mathcal{H}^s$ which are basic for the kind of iterated function systems which give Cantor constructions built on scaling and translations by lattices.

Acknowledgements

The authors gratefully acknowledge constructive discussions with Richard Gundy and Roger Nussbaum. Further, we are pleased to thank Brian Treadway for help with TEX, with style files, and with proofreading. And especially for making a number of very helpful suggestions. Both authors have benefited from discussions with the members of our NSF Focused Research Group (DMS-0139473), "Wavelets, Frames, and Operator Theory," especially from discussions with Larry Baggett, John Benedetto, David Larson, and Gestur Ólaffson. The second named author is grateful for helpful discussions and suggestions from Kathy Merrill, Wayne Lawton, Paul Muhly, Judy Packer, Iain Raeburn, and Wai-Shing Tang. Several of these were participants and organizers of an August 4–7, 2004, Wavelet Workshop ("Functional and harmonic analysis of wavelets and frames") at The National University of Singapore (NUS). And we benefited from inspiration and enlightening discussions with this workshop as well. The second named author is grateful to the hosts at NUS in Singapore for support and generous hospitality in August of 2004. Finally, we thank Daniel Alpay for inviting us to write this paper, and for encouragements.

References

[1] P. Abry, P. Flandrin, M.S. Taqqu, and D. Veitch, *Wavelets for the analysis, estimation, and synthesis of scaling data*, Self-Similar Network Traffic and Performance Evaluation (K. Park and W. Willinger, eds.), Wiley-Interscience, New York, 2000, pp. 39–88.

[2] D. Alpay, A. Dijksma, and J. Rovnyak, *A theorem of Beurling-Lax type for Hilbert spaces of functions analytic in the unit ball*, Integral Equations Operator Theory **47** (2003), no. 3, 251–274.

[3] H. Araki, *The work of Alain Connes*, Fields Medallists' lectures, World Sci. Ser. 20th Century Math., vol. 5, World Scientific, River Edge, NJ, 1997, pp. 329–336.

[4] W. Arveson, *The Dirac operator of a commuting d-tuple*, J. Funct. Anal. **189** (2002), no. 1, 53–79.

[5] ______, *Four lectures on noncommutative dynamics*, Advances in Quantum Dynamics (South Hadley, MA, 2002) (G.L. Price, B.M. Baker, P.E.T. Jorgensen, and P.S. Muhly, eds.), Contemp. Math., vol. 335, American Mathematical Society, Providence, RI, 2003, pp. 1–55.

[6] S. Avdonin and W. Moran, *Ingham-type inequalities and Riesz bases of divided differences*, Int. J. Appl. Math. Comput. Sci. **11** (2001), no. 4, 803–820, Mathematical methods of optimization and control of large-scale systems (Ekaterinburg, 2000).

[7] L.W. Baggett, P.E.T. Jorgensen, K.D. Merrill, and J.A. Packer, *Construction of Parseval wavelets from redundant filter systems*, J. Math. Phys. **46** (2005), no. 8. doi:10.1063/1.1982768

[8] ———, *An analogue of Bratteli-Jorgensen loop group actions for GMRA's*, Wavelets, Frames, and Operator Theory (College Park, MD, 2003) (C. Heil, P.E.T. Jorgensen, and D.R. Larson, eds.), Contemp. Math., vol. 345, American Mathematical Society, Providence, 2004, pp. 11–25.

[9] V. Baladi, *Positive Transfer Operators and Decay of Correlations*, World Scientific, River Edge, NJ, Singapore, 2000.

[10] J.A. Ball and V. Vinnikov, *Functional models for representations of the Cuntz algebra*, Operator Theory, Systems Theory and Scattering Theory: Multidimensional Generalizations, (D. Alpay and V. Vinnikov, eds.), Oper. Theory Adv. Appl., vol. 157, Birkhäuser, Basel, 2005, pp. 1–60.

[11] A.F. Beardon, *Iteration of Rational Functions: Complex Analytic Dynamical Systems*, Graduate Texts in Mathematics, vol. 132, Springer-Verlag, New York, 1991.

[12] J.J. Benedetto, E. Bernstein, and I. Konstantinidis, *Multiscale Riesz products and their support properties*, Acta Appl. Math. **88** (2005), no. 2, 201–227.

[13] S. Bildea, D.E. Dutkay, and G. Picioroaga, *MRA super-wavelets*, New York J. Math. **11** (2005), 1–19.

[14] A. Bisbas, *A multifractal analysis of an interesting class of measures*, Colloq. Math. **69** (1995), no. 1, 37–42.

[15] L. Blum, *Computing over the reals: Where Turing meets Newton*, Notices Amer. Math. Soc. **51** (2004), no. 9, 1024–1034.

[16] J.-B. Bost and A. Connes, *Hecke algebras, type* III *factors and phase transitions with spontaneous symmetry breaking in number theory*, Selecta Math. (N.S.) **1** (1995), 411–457.

[17] O. Bratteli and P.E.T. Jorgensen, *Isometries, shifts, Cuntz algebras and multiresolution wavelet analysis of scale N*, Integral Equations Operator Theory **28** (1997), 382–443.

[18] ———, *Iterated function systems and permutation representations of the Cuntz algebra*, Mem. Amer. Math. Soc. **139** (1999), no. 663.

[19] ———, *Wavelets through a Looking Glass: The World of the Spectrum*, Applied and Numerical Harmonic Analysis, Birkhäuser, Boston, 2002.

[20] O. Bratteli, P.E.T. Jorgensen, K.H. Kim, and F. Roush, *Decidability of the isomorphism problem for stationary AF-algebras and the associated ordered simple dimension groups*, Ergodic Theory Dynam. Systems **21** (2001), 1625–1655, corrigendum: **22** (2002), 633.

[21] ———, *Computation of isomorphism invariants for stationary dimension groups*, Ergodic Theory Dynam. Systems **22** (2002), 99–127.

[22] O. Bratteli and D.W. Robinson, *Operator Algebras and Quantum Statistical Mechanics*, 2nd ed., vol. II, Springer-Verlag, Berlin–New York, 1996.

[23] B. Brenken, and P.E.T, Jorgensen, *A family of dilation crossed product algebras*, J. Operator Theory **25**, (1991), 299–308.

[24] J. Brodzki, A. Connes, and D. Ellwood, *Polarized modules and Fredholm modules*, Mat. Fiz. Anal. Geom. **2** (1995), no. 1, 15–24.

[25] H. Brolin, *Invariant sets under iteration of rational functions*, Ark. Mat. **6** (1965), 103–144 (1965).

[26] G. Brown, *Riesz products and generalized characters*, Proc. London Math. Soc. (3) **30** (1975), 209–238.

[27] G. Brown and W. Moran, *Products of random variables and Kakutani's criterion for orthogonality of product measures*, J. London Math. Soc. (2) **10** (1975), no. part 4, 401–405.

[28] A. Cohen and R.D. Ryan, *Wavelets and Multiscale Signal Processing*, Applied Mathematics and Mathematical Computation, vol. 11, Chapman & Hall, London, 1995.

[29] Albert Cohen, *Numerical Analysis of Wavelet Methods*, Studies in Mathematics and its Applications, vol. 32, North-Holland Publishing Co., Amsterdam, 2003.

[30] R.R. Coifman, Y. Meyer, and V. Wickerhauser, *Numerical harmonic analysis*, Essays on Fourier Analysis in Honor of Elias M. Stein (Princeton, 1991) (C. Fefferman, R. Fefferman, and S. Wainger, eds.), Princeton Mathematical Series, vol. 42, Princeton University Press, Princeton, NJ, 1995, pp. 162–174.

[31] A. Connes, *Noncommutative Geometry*, Academic Press, San Diego, 1994.

[32] X. Dai and D.R. Larson, *Wandering vectors for unitary systems and orthogonal wavelets*, Mem. Amer. Math. Soc. **134** (1998), no. 640, viii+68.

[33] I. Daubechies, B. Han, A. Ron, and Z. Shen, *Framelets: MRA-based constructions of wavelet frames*, Appl. Comput. Harmon. Anal. **14** (2003), no. 1, 1–46.

[34] I. Daubechies, H.J. Landau, and Z. Landau, *Gabor time-frequency lattices and the Wexler-Raz identity*, J. Fourier Anal. Appl. **1** (1995), no. 4, 437–478.

[35] D.L. Donoho, *Renormalizing experiments for nonlinear functionals*, Festschrift for Lucien Le Cam, Springer, New York, 1997, pp. 167–181.

[36] J.L. Doob, *Application of the theory of martingales*, Le Calcul des Probabilités et ses Applications, Colloques Internationaux du Centre National de la Recherche Scientifique, no. 13, Centre National de la Recherche Scientifique, Paris, 1949, pp. 23–27.

[37] ———, *Notes on martingale theory*, Proc. 4th Berkeley Sympos. Math. Statist. and Prob., Vol. II, University of California Press, Berkeley, CA, 1961, pp. 95–102.

[38] R.G. Douglas and G. Misra, *Quasi-free resolutions of Hilbert modules*, Integral Equations Operator Theory **47** (2003), no. 4, 435–456.

[39] D.E. Dutkay, *Low-pass filters and representations of the Baumslag-Solitar group*, preprint, University of Iowa, 2004, http://www.arxiv.org/abs/math.CA/0407344. To appear in the Transactions of Amer. Math. Soc.

[40] ———, *Harmonic analysis of signed Ruelle transfer operators*, J. Math. Anal. Appl. **273** (2002), 590–617.

[41] ———, *The wavelet Galerkin operator*, J. Operator Theory **51** (2004), no. 1, 49–70.

[42] D.E. Dutkay and P.E.T. Jorgensen, *Disintegration of projective measures*, Proc. Amer. Math. Soc., to appear, http://arxiv.org/abs/math.CA/0408151.

[43] ———, *Martingales, endomorphisms, and covariant systems of operators in Hilbert space*, J. Operator Theory, to appear, http://arXiv.org/abs/math.CA/0407330.

[44] ———, *Hilbert spaces of martingales supporting certain substitution-dynamical systems*, Conform. Geom. Dyn. **9** (2005), 24–45.

[45] ———, *Wavelets on fractals*, preprint, University of Iowa, 2003, http://arXiv.org/abs/math.CA/0305443, to appear in Rev. Mat. Iberoamericana.

[46] Paul Erdös, *On the smoothness properties of a family of Bernoulli convolutions*, Amer. J. Math. **62** (1940), 180–186.

[47] R. Exel, *A new look at the crossed-product of a C^*-algebra by an endomorphism*, Ergodic Theory Dynam. Systems **23** (2003), no. 6, 1733–1750.

[48] ———, *KMS states for generalized gauge actions on Cuntz-Krieger algebras (an application of the Ruelle-Perron-Frobenius theorem)*, Bull. Braz. Math. Soc. (N.S.) **35** (2004), no. 1, 1–12.

[49] A. Fan and Y. Jiang, *On Ruelle-Perron-Frobenius operators. I. Ruelle theorem*, Comm. Math. Phys. **223** (2001), no. 1, 125–141.

[50] Alexandre Freire, Artur Lopes, and Ricardo Mañé, *An invariant measure for rational maps*, Bol. Soc. Brasil. Mat. **14** (1983), no. 1, 45–62.

[51] R.F. Gundy, *Two remarks concerning wavelets: Cohen's criterion for low-pass filters and Meyer's theorem on linear independence*, The Functional and Harmonic Analysis of Wavelets and Frames (San Antonio, 1999) (L.W. Baggett and D.R. Larson, eds.), Contemp. Math., vol. 247, American Mathematical Society, Providence, 1999, pp. 249–258.

[52] P.E.T. Jorgensen, *Analysis and Probability: Wavelets, Signals, Fractals*, Grad. Texts in Math., vol. 234, Springer-Verlag, New York, to appear 2006.

[53] ———, *Minimality of the data in wavelet filters*, Adv. Math. **159** (2001), no. 2, 143–228, with an appendix by Brian Treadway.

[54] ———, *Ruelle operators: Functions which are harmonic with respect to a transfer operator*, Mem. Amer. Math. Soc. **152** (2001), no. 720.

[55] ———, *Matrix factorizations, algorithms, wavelets*, Notices Amer. Math. Soc. **50** (2003), no. 8, 880–894, http://www.ams.org/notices/200308/200308-toc.html.

[56] J.-P. Kahane, *Sur la distribution de certaines séries aléatoires*, Colloque de Théorie des Nombres (Univ. Bordeaux, Bordeaux, 1969), Bull. Soc. Math. France, Mém. No. 25, Soc. Math. France Paris, 1971, pp. 119–122.

[57] Y. Katznelson, *Intégrales de produits de Riesz*, C. R. Acad. Sci. Paris Sér. I Math. **305** (1987), 67–69.

[58] A. Kumjian and J. Renault, *KMS states on C^*-algebras associated to expansive maps*, preprint 2003.

[59] T. Lindstrøm, *Fractional Brownian fields as integrals of white noise*, Bull. London Math. Soc. **25** (1993), no. 1, 83–88.

[60] R. Mañé, *On the uniqueness of the maximizing measure for rational maps*, Bol. Soc. Brasil. Mat. **14** (1983), no. 1, 27–43.

[61] R.D. Mauldin and M. Urbański, *Graph Directed Markov Systems: Geometry and Dynamics of Limit Sets*, Cambridge Tracts in Mathematics, vol. 148, Cambridge University Press, Cambridge, 2003.

[62] Y. Meyer, *Produits de Riesz généralisés*, Séminaire d'Analyse Harmonique, 1978–1979, Publications Mathématiques d'Orsay 79, vol. 7, Université de Paris-Sud, Faculté Des Sciences-Mathématiques, Orsay, 1979, pp. 38–48.

[63] ______, *The role of oscillations in some nonlinear problems*, European Congress of Mathematics, Vol. I (Barcelona, 2000), Progr. Math., vol. 201, Birkhäuser, Basel, 2001, pp. 75–99.

[64] Y. Meyer, F. Sellan, and M.S. Taqqu, *Wavelets, generalized white noise and fractional integration: The synthesis of fractional Brownian motion*, J. Fourier Anal. Appl. **5** (1999), no. 5, 465–494.

[65] R. Okayasu, *Type III factors arising from Cuntz–Krieger algebras*, Proc. Amer. Math. Soc. **131** (2003), no. 7, 2145–2153.

[66] C. Radin, *Global order from local sources*, Bull. Amer. Math. Soc. (N.S.) **25** (1991), 335–364.

[67] ______, *Miles of Tiles*, Student Mathematical Library, vol. 1, American Mathematical Society, Providence, 1999.

[68] M.A. Rieffel, *von Neumann algebras associated with pairs of lattices in Lie groups*, Math. Ann. **257** (1981), no. 4, 403–418.

[69] F. Riesz, *Über die Fourierkoeffizienten einer stetigen Funktion von beschränkter Schwankung*, Math. Z. **2** (1918), 312–315.

[70] G. Ritter, *On Kakutani's dichotomy theorem for infinite products of not necessarily independent functions*, Math. Ann. **239** (1979), 35–53.

[71] D. Ruelle, *Thermodynamic formalism*, Encyclopedia of Mathematics and Its Applications, vol. 5, Addison-Wesley Publishing Co., Reading, Mass., 1978.

[72] ______, *The thermodynamic formalism for expanding maps*, Comm. Math. Phys. **125** (1989), no. 2, 239–262.

[73] G. Skandalis, *Géométrie non commutative d'après Alain Connes: la notion de triplet spectral*, Gaz. Math. (2002), no. 94, 44–51.

[74] A. Volberg, and P. Yuditskii, *Noncommutative Perron-Frobenius-Ruelle theorem, two-weight Hilbert transform, and almost periodicity*, preprint, arxiv:math.SP/0408259, 2004.

[75] S.G. Williams (ed.), *Symbolic Dynamics and Its Applications*, Proceedings of Symposia in Applied Mathematics, vol. 60, American Mathematical Society, Providence, RI, 2004.

Dorin Ervin Dutkay
Department of Mathematics
Rutgers University
New Brunswick, NJ 08901, USA
e-mail: ddutkay@math.rutgers.edu

Palle E.T. Jorgensen
Department of Mathematics
The University of Iowa
Iowa City, IA 52242, USA
e-mail: jorgen@math.uiowa.edu

Operator Theory:
Advances and Applications, Vol. 167, 127–155

Noncommutative Trigonometry

Karl Gustafson

Abstract. A unified account of a noncommutative operator trigonometry originated in 1966 by this author and its further developments and applications to date will be given within a format of a historical trace. Applications to wavelet and multiscale theories are included. A viewpoint toward possible future enlargement will be fostered.

Mathematics Subject Classification (2000). 15A18, 47A05, 65N12, 65N55, 62H12, 81Q10.

Keywords. noncommutative operator trigonometry; antieigenvalue; linear algebra; quantum mechanics; statistics; numerical analysis; wavelets; multiscale systems; iterative methods.

1. Introduction

The purpose of this paper is to provide a unified account for the first time of a noncommutative trigonometry which this author initiated in 1966. This noncommutative operator trigonometry involves notions of antieigenvalues and antieigenvectors, terms which I introduced for entities which arise naturally within the theory, and which carry interest in their own right. Applications of this noncommutative trigonometry to date include, roughly and chronologically: perturbation of operator semigroups, positivity of operator products, Markov processes, Rayleigh-Ritz theory, numerical range, normal operators, convexity theory, minimum residual and conjugate gradient solvers, Richardson and relaxation schemes, wavelets, domain decomposition and multilevel methods, control theory, scattering theory, preconditioning and condition number theory, statistical estimation and efficiency, canonical correlations, Bell's inequalities, quantum spin systems, and quantum computing. Further development and applications are expected.

The outline of the paper is as follows. In Section 2, we describe the first period, 1966–1972, from which all of the key notions came. These were all due to this author. In Section 3, I describe the second period 1973 to 1993, in which my own research interests were mainly elsewhere, but during which two of my Ph.D. students, D. Rao and M. Seddighin, made contributions to the theory. In Section 4, I describe the third and most active period, 1994 to 2005, during which I applied this noncommutative trigonometry to most of the application areas described above.

The format of the paper will be chronological and historical. It is my belief that this noncommutative operator trigonometry will eventually become an established chapter of linear algebra. Thus, to that end, I want to expose its overall scope and geometrical content here. For archival purposes I have attempted to list all of my publications [1–60] on this noncommutative trigonometry to date. Very few other mathematicians have embraced this theory or contributed to it, but I will mention some in Section 5. To help the reader's overall perspective and interest, I will at times add some commentary, even of a personal nature. Thus the style of this account will not discriminate between grammatical person, i.e., I, we, one, this author, and will just use whichever seems appropriate within local context.

Full detail would require writing a book. In fact we have published two books [26] and [29] which contain significant chapters dealing with this noncommutative operator trigonometry. But both of those books were written in 1995 and there has been considerable new development and interesting applications in the ten years since then. Nonetheless I will often refer to those books when possible to keep the exposition here within space and time bounds.

Finally, it is my hope that by providing this unified account, I have fostered a viewpoint which will enable future enlargement of this theory and its applications. In particular, from time to time I will remind or emphasize that this theory, from its beginning, has been: noncommutative. Of course it contains the commutative case, but it is much more valuable for noncommuting operators. To me, operator theory has always been noncommutative. Operators A and B should always be first approached as not commuting with each other. In most engineering applications, there would be no dynamics and the structures would all fall down were their governing linear systems commutative. In a metaphysical sense, then, I would like this paper to provide a take-off point into a more abstract conceptual frame somewhat akin to that of noncommutative geometry. Section 6 contains a few preliminary results and remarks toward that goal.

2. The first (active) period 1966–1972

In [1], my first published mathematics paper, I contributed to the perturbation theory of linear operators. One of the important theorems there is one originally found in Hilbert space independently by T. Kato and F. Rellich. Here is a Banach space version. Let $G(1,0)$ denote the densely defined infinitesimal generators A of a contraction semigroup Z_t on a Banach space X. Additively perturb A by a regular perturbation B, i.e., domain $D(B) \supset D(A)$, which is relatively small: $||Bx|| \leqq a||x|| + b||Ax||$ for some $b < 1$. Then $A + B$ is also in $G(1,0)$, i.e., $A + B$

generates a contraction semigroup, iff $A + B$ is dissipative: Re $[(A + B)x, x] \leqq 0$ for all x in $D(A)$. Here $[y, x]$ denotes any semi-inner-product on the Banach space.

I wanted a general multiplicative perturbation theorem and obtained the following: see [2]–[7] for details and related developments. Let A be a generator in $G(1, 0)$ and let B be bounded on X and strongly accretive: Re $[Bx, x] \geqq m||x||^2$ for all $x, m > 0$. Then BA is in $G(1, 0)$, i.e., BA generates a contraction semigroup, iff BA is dissipative: Re $[BAx, x] \leqq 0$ for all x in $D(A)$.

Thus I arrived in 1966 at the core issue from which the noncommutative operator trigonometry to be exposed in this paper arose: when is dissipativity preserved in an operator composition of an accretive operator B with a dissipative operator A? It occurred to me to try to answer this question in a way similar to that by which I had proved the rest of the just stated multiplicative perturbation theorem: reduce it to the Kato–Rellich additive perturbation theorem. Inserting a free parameter $\epsilon > 0$, which will not alter whether or not ϵBA is a generator or not, one writes $\epsilon BA = A + (\epsilon B - I)A$ and one notes that $||(\epsilon B - I)Ax|| \leqq ||\epsilon B - I|| ||Ax||$ implies by the Kato–Rellich Theorem that $(\epsilon B - I)A$ is an acceptable additive perturbation provided that $b \equiv ||\epsilon B - I|| < 1$ for some $\epsilon > 0$. Indeed one finds that the norm curve $||\epsilon B - I||$ dips below 1 for an interval $(0, \epsilon_c)$ whenever B is strongly accretive. Applying the same additive decomposition to the multiplicative perturbation in the semi-inner-product, one finds

$$\begin{aligned} \mathrm{Re}[\epsilon BAx, x] &= \mathrm{Re}[\epsilon B - I)Ax, x] + \mathrm{Re}[Ax, x] \\ &\leqq ||\epsilon B - I|| ||Ax|| ||x|| + \mathrm{Re}[Ax, x] \end{aligned} \tag{2.1}$$

from which one obtains the sufficient condition for ϵBA, hence BA, to be dissipative:

$$\min_{\epsilon > 0} ||\epsilon B - I|| \leqq \inf_{\substack{x \in D(A) \\ x \neq 0}} \mathrm{Re} \frac{[(-A)x, x]}{||Ax|| ||x||}. \tag{2.2}$$

As A being dissipative is the same as $-A$ being accretive, it is easier to repose the result (2.2) as follows, and for simplicity, let us now just restrict attention to a Hilbert space X and its usual inner product. Define for accretive A

$$\mu_1(A) \equiv \inf_{\substack{x \in D(A) \\ x \neq 0}} \mathrm{Re} \frac{\langle Ax, x \rangle}{||Ax|| ||x||} \equiv \cos \phi(A) \tag{2.3}$$

and let $\nu_1(B)$ denote $\min_\epsilon ||\epsilon B - I||$ for accretive B. Then when $\nu_1(B) \leqq \mu_1(A)$, BA will be accretive. The intuition for calling $\mu_1(A)$ a $\cos \phi(A)$ is rather immediate. On the other hand, I puzzled on the entity $\nu_1(B)$ for awhile and decided that it too must be trigonometric. This I showed in 1968 in [6] by a min-max argument, see also [11]. The result is that one can define for bounded accretive A and $-\infty < \epsilon < \infty$ the entity

$$\nu_1(A) \equiv \min_{\epsilon} ||\epsilon A - I|| \equiv \sin \phi(A). \tag{2.4}$$

The result of [6] is then: $\cos^2\phi(A)+\sin^2\phi(A)=1$. This important result gave me an operator trigonometry.

I emphasize that one does not have this operator (noncommutative) trigonometry until one has $\sin\phi(A)$. One needs both a cosine and a sine to develop a trigonometry. I also emphasize that although of course these results contain the commutative cases too, they are not as interesting. See [3], [4], [5] for how commutative situations often become trivial or may be resolved by other means.

Thus from (2.2) the condition for the multiplicative perturbation BA to generate a contraction semigroup became

$$\sin\phi(B) \overset{<}{=} \cos\phi(A). \tag{2.5}$$

However, I was a little disappointed that my theory was now restricting itself to bounded B (I like unbounded operators). So Gunter Lumer and I extended results (but not the trigonometry) to include unbounded B [12]. Bruce Calvert and I obtained results for nonlinear semigroups [13]. Ken-iti Sato and I extended both additive and multiplicative perturbation results to positivity preserving (e.g., real-valued Markov) semigroups [9]. These papers contained no trigonometry.

In [8] Bruno Zwahlen and I showed that all unbounded densely defined accretive operators A in a Hilbert space, have $\cos\phi(A)=0$. This finding, although of interest in itself in distinguishing unbounded and bounded operators geometrically, really disappointed me because it meant that the perturbation criteria (2.2) was not useful when the semigroup generator A was unbounded in a Hilbert space.

However, the trigonometry itself seemed interesting, so when invited to the UCLA Symposium on Inequalities in 1969, see the book of [11], I presented the main results there. I decided to call $\mu_1(A)$ the first antieigenvalue of A because it measured the greatest turning angle of A: just the opposite geometrical effect from that of an eigenvalue, which measures the stretching effect of a vector which is turned not at all by A. Real antieigenvalues, Imaginary antieigenvalues, Total antieigenvalues, and Higher antieigenvalues were defined, and the min-max result $\cos^2\phi(A)+\sin^2\phi(A)=1$ exposed. Only mentioned in [11] but not detailed was the Euler equation

$$2||Ax||^2||x||^2(\mathrm{Re}A)x-||x||^2\mathrm{Re}\langle Ax,x\rangle A^*Ax-||Ax||^2\mathrm{Re}\langle Ax,x\rangle x=0 \tag{2.6}$$

which is the functional derivative of (2.3) and which must be satisfied by any first antieigenvector. This equation is also satisfied by all eigenvectors of A when A is selfadjoint or normal. Thus (2.6) may be regarded as a significant extension of the Rayleigh-Ritz variational theory for eigenvalues. For A a symmetric $n\times n$ positive definite matrix, I established that

$$\cos\phi(A)=\frac{2\sqrt{\lambda_1\lambda_n}}{\lambda_1+\lambda_n},\quad \sin\phi(A)=\frac{\lambda_n-\lambda_1}{\lambda_n+\lambda_1} \tag{2.7}$$

and that the first antieigenvectors are the pair

$$x_\pm=\pm\left(\frac{\lambda_n}{\lambda_1+\lambda_n}\right)^{1/2}x_1+\left(\frac{\lambda_1}{\lambda_1+\lambda_n}\right)^{1/2}x_n \tag{2.8}$$

where λ_1 is the smallest eigenvalue, with normalized eigenvector x_1, and λ_n is the largest eigenvalue, with normalized eigenvector x_n.

To fix ideas, let $A = \begin{bmatrix} 9 & 0 \\ 0 & 16 \end{bmatrix}$, then by (2.7) and (2.8) we have the operator turning angle $\phi(A) = 16.2602°$ and the two antieigenvectors (not normalized here) $x_+ = (4, 3)$ and $x_- = (-4, 3)$.

The papers [1]-[13] constitute the initial period during which I originated this theory of antieigenvalues and antieigenvectors and operator turning angles. In retrospect, I was fortunate that my motivating original question, that of multiplicative perturbation of semigroup generators, brought forth both $\cos\phi(A)$ and $\sin\phi(A)$, of necessity. And in that application, the operator compositions BA are thoroughly noncommutative.

Commentary. One will notice common ingredients of the important min-max result [6], [11] proof in my proof [10] of the Toeplitz-Hausdorff Theorem for the convexity of the numerical range $W(A)$. That is why I include [10] in this list [1]–[60].

3. The second (intermittent) period 1973–1993

A year sabbatical 1971–72 to the Theoretical Physics Institute at the University of Geneva and the Ecole Polytechnique Fédérale in Lausanne, where the research interests were principally quantum scattering theory and bifurcation theory, respectively, brought to an end my initial contributions to the operator trigonometry. However, my Ph.D. student Duggirala Rao found a connection to a 1969 paper by the noted Russian mathematician M.G. Krein. See my survey [25] for more details about Krein's work. Also together Rao and I looked at the operator trigonometry more generally within the context of the numerical range $W(AB)$. From this resulted the 1977 paper [14]. It should be noted that most of the issues dealt with in that paper are thoroughly noncommutative. In particular, some generalizations of the sufficient condition (2.5) were obtained. Also the accretivity of the product T^2 (admittedly commutative, but for an unbounded operator T) was considered within the framework of sesquilinear forms. Also I wrote an invited paper [15] related to [1] in 1983. The main concerns there centered on domains of perturbed unbounded operators.

Then with Ph.D. student Morteza Seddighin, we extended related 1980 results of Chandler Davis, with emphasis on the Re and Total antieigenvalue theory for normal operators. This resulted in the two papers [16] and [18] in 1989 and 1993, respectively. I will elaborate a bit on these developments in Section 5. Recently Seddighin and I returned [59] to this theory and resolved some shortcomings in [16], [18] and also some incomplete statements in Davis' analysis. In particular we completely resolved an issue which I called "possible zero denominators" which arose from our use of LaGrange multiplier techniques in [16], [18].

An important event of interest which occurred during this rather inactive (as concerns the operator trigonometry) period, was an invitation to speak at a conference [17] on convexity in Dubrovnik, Yugoslavia, in June 1990. The organizers especially asked me to speak about my semigroup and antieigenvalue results from 20 years earlier! This I did. Relately, although not immediately obvious, in 1978 several engineering departments had asked me to offer courses on computational partial differential equations, especially computational fluid dynamics. This I did, from 1978 to 1990. One learns working in that field that 70% of supercomputing time is spent on linear solvers. In particular, in such research I had used some steepest descent and conjugate gradient scheme computations and had learned of the basic convergence rate bound

$$E(x_{k+1}) \leqq \left(1 - \frac{4\lambda_1\lambda_n}{(\lambda_1 + \lambda_n)^2}\right) E(x_k). \tag{3.1}$$

Here $E(x)$ is the error $\langle (x - x^*), A(x - x^*)\rangle/2$, x^* the true solution. This bound also occurs in a key way in some parts of optimization theory. Immediately from my operator trigonometric expressions (2.7) I knew that what (3.1) really meant, geometrically, was trigonometric:

$$E(x_{k+1}) \leqq \sin^2 \phi(A) E(x_k). \tag{3.2}$$

Although (3.1) was known as a form of the Kantorovich inequality, nowhere in the numerical nor in the optimization literature did I find this important geometrical result: that gradient and conjugate gradient convergence rates are fundamentally trigonometric. That is, they are a direct reflection of the elemental fact of A's maximum turning angle $\phi(A)$. The known conjugate gradient convergence rate, by the way, becomes, trigonometrically written now,

$$E(x_k) \leqq 4(\sin \phi(A^{1/2}))^{2k} E(x_0). \tag{3.3}$$

See [19] where I published this result (3.3) for conjugate gradient convergence. And I first announced the steepest descent result (3.2) at the 1990 Dubrovnik conference [17].

Summarizing, the period 1973–1993 was one of low, almost nonexistent, priority for the operator trigonometry. Countervailingly, I was very occupied with learning some computational linear algebra. In addition to a significant number of publications with four Ph.D. students which I produced in the departments of Aerospace Engineering, Civil Engineering, Chemical Engineering, and Electrical Engineering, I had also been asked to be the only mathematician in a seven department successful attempt to bring a $22 million Optoelectronic Computing Systems Engineering Research Center to the University of Colorado. I was in charge of algorithms, especially those to be implemented on a large optoelectronic neural network. This activity took most of my research time in the interval 1988–1993.

But what I learned from this, essentially engineering, research in this period 1978–1993, accrued to produce later dividends for the operator trigonometry, as I will discuss in the next Section. Furthermore, I had worked with physicists at the

Solvay Institute of Professor Ilya Prigogine in Brussels during the period 1978–2000, and what I learned there also was indirectly of great importance later for the operator trigonometry, as I will recount later.

Commentary. There are three interesting and related vignettes coming out of the Dubrovnik meeting in June 1990.

First, as I flew in from Vienna, sitting beside me was a lovely coed of a joint American–Yugoslavian family, returning from UCLA. She whispered detailed predictions to me of a coming war, and why. I didn't know if I should believe her. But the guards who rather roughly shook us down as we departed the airplane at the Dubrovnik airport, added credence to her admonitions. This was several months before the war actually broke out. Where was the U.S. intelligence community?

Second, I knew that my discovery of the fundamental direct trigonometric meaning of the Kantorovich inequality, viz., as it occurs in (3.1)–(3.3), would be of general importance. So I intentionally put it into the proceedings of the Dubrovnik meeting. Then I waited. War broke out. I may have been the only meeting participant who truly was concerned as to whether the Proceedings would in fact appear. Somehow, the editors (Časlav Stanojević and Olga Hadžić) did manage to publish a Proceedings, and hence my contribution [17], out of Novi Sad, Yugoslavia.

Third, I submitted the paper [19] in 1993, a paper essentially invited by the organizers of a 1992 numerical range conference to which I had been unable to go. In the manuscript [19] I cited the paper [17] in the Yugoslavian meeting Proceedings, and its result. Certainly those proceedings were not very accessible to the general public. A couple of months later I received a very confidential fax from a rather secretive-sounding commercial agency, wanting a copy of my Yugoslavian paper, citing it as such, and offering to pay me a sum ($25 or $65, I cannot remember) for the paper. I was to send the copy of my paper to the commercial firm's address. They said they were getting the paper from me for some client who was to remain unnamed. This is probably the strangest "preprint request" I have ever received. In any case, I did not comply.

4. The third (most active) period 1994–2005

Following a disastrous (in my opinion) and illegal (in my opinion) split of our mathematics department into separate applied mathematics and pure mathematics in 1990, I had to decide which to join. Caught directly in the middle and with one foot in each subject, I chose (for moral and legality reasons, in my opinion) to stay in the pure mathematics department. Within two years it became clear to me that I had been "closed out" of any future meaningful applied research teams and applied Ph.D. students and applied grants for work of the kind I described in the section above. A second happening was the retirement conference here for my very good friend and matrix theory expert, John Maybee (now deceased). Similar to the conference volume of LMA in which [19] appeared, a LAA conference volume for John was planned. Having just gone back to the operator trigonometry in [17] and [19], I decided to specialize the operator trigonometry to matrix trigonometry [22] in the paper for John. Simultaneously there was a retirement volume of LAA for Chandler Davis, and I put more antieigenvalue theory into that paper [20]. These three papers [19], [20], [22] signalled a return to the operator trigonometry as a higher priority.

Meanwhile, I presented my new operator-trigonometric results [21] for certain iterative linear solvers, at a 1994 computational linear algebra conference here in Colorado. I and my students had presented several multigrid solvers research applied to computational fluid dynamics at these famous Copper Mountain–Breckenridge numerical conferences earlier in the 1980's. But my presentation in 1994 received a cold shoulder. My impression was that no one there understood my theory. This impression has been enhanced, by the way, at some other conferences of the same community. However, I was then invited to the Nijmegen 1996 numerical conference and my results were published there [24], [27].

Moreover, at about the same time I had been invited by Hans Schneider to contribute to the Wielandt volume [25]. It was Schneider who pointed out Wielandt's (partial) operator trigonometry to me, of which I had been unaware. See my discussion in [25]. Also I had given six invited lectures in Japan in 1995. These resulted in the 1996 book [23] which then became the wider-distributed 1997 book [29]. Also Duggirala Rao and I had decided in 1990 to write a book on Numerical Range. That book [26] appeared in 1997. Both of those books contain accounts of the operator trigonometry. As well, the Encyclopedia [28] sought a description of antieigenvalue theory.

In summary, due to the (political) defaulting of me out of research sponsored by the applied mathematics community, plus the welcoming of me by the more mathematical linear algebra community, I was able to accelerate research on my operator trigonometry, which became of benefit to both communities! The transition period is reflected in the publications [19]–[30] from 1994 into 1998. What are some of the main results?

I have already mentioned in Section 3 above the convergence ratio for steepest descent and conjugate gradient solvers above, which came earlier. An important subsequent paper is [27]. There, in particular, I show that the very basic Richardson scheme, is trigonometric. That had never been realized before. In more detail, the Richardson iteration can be written as $x_{k+1} = x_k + \alpha(b - Ax_k)$ and I showed that

$$\alpha_{\text{optimal}} = \epsilon_m = \text{Re}\frac{\langle Ax_+, x_+\rangle}{||Ax_+||^2}, \quad \rho_{\text{optimal}} = \sin\phi(A) \tag{4.1}$$

where α_{optimal} is the optimal (relaxation) choice and where ρ_{optimal} is the optimal convergence rate. In (4.1) ϵ_m is the operator-trigonometric value that gives you $\sin\phi(A)$ in (2.4), and x_+ denotes either of A's antieigenvectors $x_\pm$. Beyond this fundamental albeit basic Richardson scheme, in [27] I also show that the known convergence rates for the more sophisticated schemes GCR, PCG, Orthomin, CGN,

CGS, BCG, GM–RES, are all to some extent, operator trigonometric. I also obtained in [27] the first geometrical (read: trigonometric) understandings of the convergence of the schemes Jacobi, SOR, SSOR, Uzawa, Chebyshev algorithms.

In addition, at the conference of [24], I was introduced to the AMLI (algebraic multilevel iteration) theory, and I was able to quickly establish connections to my operator trigonometry. This is reported in [27] and here is a summary. Consider a nested sequence of finite element meshes

$$\Omega_\ell \supset \Omega_{\ell-1} \supset \cdots \supset \Omega_{k_0} \tag{4.2}$$

and let $V_\ell, V_{\ell-1}, \ldots, V_{k_0}$ be corresponding finite element spaces for a finite element method applied to a second order elliptic boundary value problem. In particular, for a chosen discretization level ℓ, let V_1 and V_2 be finite element subspaces of V_ℓ such that $V_1 \cap V_2 = 0$. Let $a(u, v)$ be the symmetric positive definite energy inner product on the Sobolev space $H^1(\Omega)$ corresponding to the given elliptic operator.

Let

$$\gamma = \sup_{u \in V_1, v \in V_2} \frac{a(u,v)}{(a(u,v) \cdot a(v,v))^{1/2}} \tag{4.3}$$

be the resulting C.B.S. (Cauchy–Bunyakowsky–Schwarz) constant. For hierarchical finite element basis functions one uses a finite element symmetric-matrix partitioning

$$A = \begin{bmatrix} A_{11} & A_{12} \\ A_{21} & A_{22} \end{bmatrix} \tag{4.4}$$

which has the property that the spectral radius $\rho(A_{11}^{-1/2} A_{12} A_{22}^{-1/2})$ equals the C.B.S. constant (4.3). In the AMLI schemes one then likes to "precondition" by D^{-1} where D is the block diagonal portion of A in (4.4). From these and other considerations I was able to show the following result: $\gamma = \sin\phi(D^{-1}A)$, and the AMLI interlevel amplification factors are all fundamentally trigonometric.

In a following paper [36], I worked out many details of the operator trigonometry of these iterative methods for the so-called model problem: the Dirichlet boundary value problem on the unit square in two dimensions. I refer the reader to [36] for details. In particular, there is a powerful scheme, especially valuable and efficient for the numerical solution of parabolic PDE problems but also competitive for elliptic problems, called ADI (alternating-direction-implicit). I was able to reveal in [36] that a beautiful sequence of operator sub-angles underlies and explains the dramatic speed-up of ADI when it is employed in its semi-iterative formulation: the optimizing parameter is allowed to change with each iteration. Then the convergence rate bounds possess an interesting recursive relationship $d(\alpha, \beta, 2q) = d\left(\sqrt{\alpha\beta}, \frac{\alpha+\beta}{2}, q\right)$.

From this [36] one obtains the trigonometric convergence rate bounds

$$
\begin{aligned}
d(\alpha, \beta, 1) \quad &= \sin \phi(A_h^{1/2}), \\
d(\alpha, \beta, 2) &= \tfrac{1-(\cos \phi(A_h))^{1/2}}{1+(\cos \phi(A_h))^{1/2}}, \\
d(\alpha, \beta, 4) &= \frac{1-\frac{\cos \phi(A_h))^{1/4}}{\cos(\phi(A_h)/2)}}{1+\frac{(\cos \phi(A_h))^{1/4}}{\cos(\phi(A_h)/2)}}.
\end{aligned}
\tag{4.5}
$$

For the example given in [36], these convergence (error-reduction) rates become $d(1) = 0.268$, $d(2) = 0.036$, $d(4) = 0.00065$, clearly illustrating the dramatic speedup of semi-iterative ADI. It should be mentioned that ADI in various guises still plays a major role in NASA aerodynamic simulation codes.

In the domain decomposition paper [30], I showed that the optimal convergence rate of the additive Schwarz alternating domain decomposition is trigonometric: specifically,

$$
\rho(M_{\theta_{\text{optimal}}}^{\text{odd } SI}) = \sin \phi(W^{-1/2} A W^{-1/2}). \tag{4.6}
$$

Here W is a weighting matrix. See [30] for related results. I note here that full proofs are given in the later paper [50]. It should be mentioned that domain decomposition methods are explicitly multiscale methods and always of inherent multilevel philosophy. Thus there are strong links between wavelet methods, multilevel methods, and multiscale and multigrid computing.

For example, in the wavelet paper [31] I showed that the wavelet reconstruction algorithm is really the Richardson iteration that I mentioned above. To see that, let a function x have wavelet frame coefficients $\{\langle x, x_n \rangle\}$ where $\{x_n\}$ is a frame basis for the Hilbert space H: there exist positive constants A and B such that for any x in H

$$
A||x||^2 \leqq \sum_n |\langle x, x_n \rangle|^2 \leqq B||x||^2. \tag{4.7}
$$

The corresponding unique frame operator $S = \Sigma_n \langle x, x_n \rangle x_n$ is symmetric positive definite with lower and upper quadratic form bounds A and B, respectively. We may write $Sx = \Sigma_n \langle x, x_n \rangle x_n = F^* F x$ and this leads to the well known reconstruction formula

$$
x = \frac{2}{A+B} \Sigma_n \langle x, x_n \rangle x_n + Rx \tag{4.8}
$$

where the error operator $R = I - \frac{2}{A+B} S$ has norm $||R|| \leqq \frac{B-A}{B+A} < 1$. The wavelet reconstruction algorithm then becomes

$$
x^{N+1} = x^N + \frac{2}{A+B}(Sx - Sx^N). \tag{4.9}
$$

As I showed in [31], this is exactly the old Richardson scheme $x^{N+1} = x^N + \alpha(b - Sx^N)$ with iteration matrix $R_\alpha = I - \alpha S$, $b = Sx$, and $\alpha = 2/(A+B)$. From my earlier computational trigonometric results, this means α is optimal and produces the optimal convergence rate $\sin \phi(S)$. When the frame bounds A and B are sharp, i.e., A and B are the smallest and largest eigenvalues of S or at least the smallest and largest values of the spectrum of S, then $\alpha = \langle Sx_+, x_+\rangle/||Sx_+||^2$ where x_+ denotes either of the two antieigenvectors $x_\pm$ of the frame operator S. Note that in [31] I have also for the first time introduced the operator angle $\phi(S)$ and its operator trigonometry for arbitrary wavelet frame operators.

I return now to the application of the operator trigonometry to computational linear algebra, if we may jump ahead here, specifically to [50], with related papers [55], [56] which I will not discuss here. But [50] is an important numerical paper which very clearly brings out the noncommutativity inherent in many of these computational linear algebra considerations. In [50] I obtain further results for domain decomposition algorithms, multigrid schemes, sparse approximate inverse methods, SOR (successive overrelaxation), and minimum residual schemes. But I may emphasize my point about noncommutativity by the results in [50] directed at preconditioning methods. In short: to solve $Ax = b$ for large systems, one tries to reduce its condition number $\kappa = \lambda_n/\lambda_1$ by premultiplying by a B which somehow resembles A^{-1}. Then one numerically solves $BAx = Bb$. Some preconditioning methods allow B to commute with A but generally that will not be the case. For A and B symmetric positive definite I show in [50] that

$$\min_{\epsilon>0} ||I - \epsilon BA||_C = \sin\phi(BA), \tag{4.10}$$

where C denotes any symmetric positive definite operator for which BA is symmetric positive definite in the C inner product $\langle Cx, y\rangle$. The left side of (4.10) is an optimal convergence rate which occurs in many preconditioning and multilevel schemes.

Another interesting result of [50] is that I reveal therein a very rich operator trigonometry which has heretofore lain unknown but which fundamentally underlies SOR schemes. The full implications of this discovery are not yet worked out. But see the Commentary at the end of this Section.

Returning to the list [1]–[60], let us now turn away from computational linear algebra. Briefly, in [33] the noncommutative trigonometry is applied to control theory. However, the control theory community seems still far ahead. The papers [34], [39] respond to a question of the matrix analysis community. In [38] I clarify the relationship of my operator angles to other angles defined in terms of the condition number κ. The paper [40] was an invited paper for the Russian audience. The paper [45] extends the operator trigonometry to arbitrary matrices A.

Next we turn to physics. In [32] I relate multiplicative perturbations and irreversibility and scattering theory within a framework of the operator trigonometry. The somewhat related papers [35] and [43] will not be discussed here. More interesting is my rather fundamental finding in 1997 that the Bell inequalities of quantum mechanics can all be embedded and treated within my operator trigonometry. This finding was first announced in [37] and further developments may be found in [42], [44], [46], [47], [49], [51], [52], [53], [60]. I think there is too much there to try to summarize here. I am continuing research on this subject.

I always felt that the operator trigonometry would have useful application to certain areas of statistics. As a result of my being sent in 1999 two papers from two journals dealing with certain criteria for best least squares estimation, to referee, I saw the connection of the operative trigonometry to statistics and submitted [41] to one of those journals. To my knowledge, [41] was the first paper to make the connection, and also the first paper to formulate the notion of antieigenmatrix. The paper was rejected. But I then published it in [48]. My results, now well thought of within the statistics community, contain the following. There was a famous conjecture of the noted mathematical statistician Durbin, which was open for the twenty-year interval 1955–1975. The conjecture was that the so-called least squares (relative) efficiency (see [48] for details)

$$\mathrm{RE}(\hat{\beta}) = \frac{|\operatorname{Cov}\beta^*|}{|\operatorname{Cov}\hat{\beta}|} = \frac{1}{|X'VX||X'V^{-1}X|} \tag{4.11}$$

of estimators, should have (sharp) lower bound

$$\mathrm{RE}(\hat{\beta}) \geqq \prod_{i=1}^{p} \frac{4\lambda_i\lambda_{n-i+1}}{(\lambda_i + \lambda_{n-i+1})^2} \tag{4.12}$$

where the λ's are the eigenvalues of the covariance matrix V. In (4.11) the β^* is the best possible estimator and X is an $n \times p$ regression design matrix. Coincidentally, I had recently [45] extended my operator trigonometry to arbitrary matrices and also in [45] I stressed (also earlier, in [22]) the alternate definition of higher antieigenvalues and their corresponding higher antieigenvectors as defined combinatorially in accordance with (2.7) but with successively smaller operator turning angles now defined by corresponding pairs of eigenvectors. This is in my opinion better than my original formulation [11] which was more in analogy with the Rayleigh–Ritz philosophy. In any case, it follows readily that the geometric meaning of statistical efficiency (4.12) then becomes clear:

$$\mathrm{RE}(\hat{\beta}) \geqq \prod_{i=1}^{p} \cos^2\phi_i(V). \tag{4.13}$$

It is that of operator turning angles.

Other intimate connections between statistical efficiency, statistical estimation, and canonical correlations may be found in [41], [48]. For example, Lagrange multiplier techniques used in statistics lead to equations for columns x_i of the regression design matrices X to satisfy

$$\frac{V^2x_i}{\langle Vx_i, x_i\rangle} + \frac{x_i}{\langle V^{-1}x_i, x_i\rangle} = 2Vx_i, \quad i = 1, \ldots, p. \tag{4.14}$$

I call this equation the Inefficiency Equation. In the antieigenvalue theory, the Euler equation (2.6) becomes in the case when A is symmetric positive definite,

$$\frac{A^2x}{\langle A^2x, x\rangle} - \frac{2Ax}{\langle Ax, x\rangle} + x = 0. \tag{4.15}$$

The combined result is the following [41], [48]. All eigenvectors x_j satisfy both the Inefficiency Equation and the Euler Equation. The only other (normalized) vectors satisfying the Inefficiency Equation are the "inefficiency vectors" (my notation)

$$x_{\pm}^{j+k} = \pm\frac{1}{\sqrt{2}}x_j + \frac{1}{\sqrt{2}}x_k. \tag{4.16}$$

The only other vectors satisfying the Euler equation are the antieigenvectors

$$x_{\pm}^{jk} = \pm\left(\frac{\lambda_k}{\lambda_j+\lambda_k}\right)^{1/2} x_j + \left(\frac{\lambda_j}{\lambda_j+\lambda_k}\right)^{1/2} x_k. \tag{4.17}$$

Further recent results along these lines may be found in [48] and [58].

To conclude this section's account, I returned [57] to some unfinished business from [11] and brought results of G. Strang on Jordan products and J. Stampfli on derivations into the operator trigonometry.

Commentary. The one paper in the list [1]–[60] to which I have not yet referred is [54], a short three page summary of a paper I presented to the computational linear algebra community at a nice conference in Napa, California in October 2003. In particular, and with respect to my comment above about a rich operator trigonometry underlying the SOR algorithm (which is still a good algorithm, by the way), I had found from an entirely different direction the following result. If one factors an arbitrary invertible $n \times n$ matrix A into its polar representation $A = QH$ where Q is unitary and H is Hermitian, then one may define the absolute condition number $C(H)$ as follows:

$$C(H) = \lim_{\delta\to 0} \sup_{\substack{\Delta A\in R^{n\times n}\\ ||\Delta A||\leqq\delta}} \frac{||H(A) - H(A+\Delta A)||_F}{\delta}. \tag{4.18}$$

In (4.18) the F denotes the Frobenius norm $(\Sigma|a_{ij}|^2)^{1/2}$ of a matrix. It is recently known (see the next section) that $C(H)$ has the exact value

$$2^{1/2}(1+\kappa^2(A))^{1/2}/(1+\kappa(A)) \tag{4.19}$$

where here κ denotes the condition number σ_1/σ_n where $\sigma_1 \geqq \cdots \geqq \sigma_n$ are the singular values of A, i.e., the positive eigenvalues of H. It is relatively straight forward to see that the operator trigonometric meaning of this condition number expression is exactly

$$C^2(H) = 1 + \sin^2\phi(H). \tag{4.20}$$

From this one sees immediately that the geometrical content of the absolute condition number (4.18) is entirely trigonometric. It is all a matter of how far H can turn a vector.

A number of other interesting relationships are found in the full paper which I announced in [54]. One is the following connection to SOR. Here I assume the additional conditions on A needed by the SOR theory have been met. Then one obtains, using results from [50], that

$$C^2(H) = \omega_b^{SOR}(H^2) \tag{4.21}$$

where ω_b is the optimal relaxation parameter for SOR.

Apparently this full paper from [54] is too theoretical for the computational linear algebra community and it rests, so far, unpublished. Perhaps the most interesting result from my analysis therein concerns the notion of antieigenmatrix, a concept I first introduced in the statistics paper [41]. Therefore, in Section 6 here I will construct an antieigenmatrix which achieves the $C(H)$ limit in (4.18), i.e., which attains the bounds (4.19) and (4.20).

5. Related work: Discussion

I have already mentioned above the early independent work of M.G. Krein [61] and Wielandt [62], both of whom however failed to see the important entity $\min_\epsilon ||\epsilon A - I||$ and thus missed $\sin\phi(A)$ and thus were not led to an operator trigonometry. My 1968 min-max theorem [6] was extended in 1971 by Asplund and Ptak [63] and because of their result (general failure of the min-max theorem for Banach spaces), I did not further pursue the Banach space context, even though I had mentioned it in 1969, see [11], p. 116. Later congenial discussions with V. Ptak confirmed to me that he did not know of my 1968 publication [6], but I had intimately discussed my min-max result with E. Asplund, who was quite excited about it, at the 1969 UCLA meetings from which finally came the 1972 book in which my antieigenvalues paper [11] appeared. In 1971 P. Hess [64] extended our result [8] that $\cos\phi(A) = 0$ for unbounded accretive operators. As we observed in [16], that need no longer be the case in certain Banach spaces.

But my first encounter with really relevant outside interest in my operator trigonometry theory came from Chandler Davis [65] who invited me to Toronto in 1981 to discuss the subject. Through Davis I also learned of the results by Mirman [66] for the calculation of antieigenvalues. I also learned for the first time that in my 1960's work I had rediscovered Kantorovich's inequality [67]. However, as I have emphasized elsewhere in this account: no-one, including Kantorovich, had seen the natural operator angle which I had found in my equivalent formulations of the inequality. As mentioned earlier, the reader may see [20] for further relationships to Davis' theory of shells of Hilbert space operators. For a recent survey of many inequalities as they relate to Kantorovich's inequality and to other issues in statistics, see [68].

Following my 1999 paper [41], which first established the connection between my operator trigonometry and statistics, and which circulated but was rejected by the journal Metrika but which I then successfully published in 2002 in [48], and perhaps due to the authors of [68] including in their widely circulated 1999 drafts of [68] some explicit mention of my antieigenvalue theory, Khatree [69], [70], [71] published three papers somewhat parallel to [41], [48] but more from the statistics viewpoint. There had also been a little earlier interest in my antieigenvalue theory by mathematicians in India, see [72]. At the 14th IWMS Conference in Auckland in March 2005, I enjoyed thoroughly agreeable discussions with the noted mathematical statistician C.R. Rao, who has fully embraced [73] my operator trigonometry as it applies to various considerations in statistics.

The same cannot be said of the computational linear algebra community. When I first presented my application of the operator trigonometry to convergence rates for iterative linear solvers at the Colorado Conference on Iterative Methods at Breckenridge, Colorado in April 1994, a colleague from Texas who knew and appreciated some of my earlier work [74] on finite element methods, quietly took me aside and gave me his opinion that no-one there would understand nor appreciate this noncommutative operator trigonometry as it applied to their matrices. Indeed, my full paper [21] was rejected by a SIAM journal from the conference. I then published my results in [24] and [27]. At the 1996 conference of [24] a very well known computational linear algebraist from Germany at my presentation asked me "but can you produce a better algorithm?" Someone else, more supportive, later pointed out to me that my questioner, although a great authority and author of several excellent books on computational linear algebra, had never himself "produced a better algorithm." Nevertheless, I do respect that viewpoint in that field of scientific endeavor: producing a better algorithm is the name of the game. It is more important than a better theory, which in my opinion, I have indeed produced.

So to my knowledge, I know of no work on this noncommutative trigonometry by others within the computational matrix algebra community. However, some, e.g., Martin Gutknecht in Zurich and Owe Axelsson in Nijmegen and Michele Benzi in Atlanta, among others, are genuinely appreciative of the theoretical contribution my geometrical viewpoint makes to the overall theory of iterative and multilevel methods. Some day, one of us may use this noncommutative matrix trigonometry to produce a faster or more accurate algorithm.

Within the quantum mechanics community, there has been great interest in my successful embedding of much of the Bell inequality theory into my noncommutative trigonometry. This has been clearly evidenced when I presented my recent results at many top-level conferences on the foundations of quantum mechanics in Europe and in the USA. However, to my knowledge, no-one and no group has pushed my results further. One reason for that is there are so many other important issues surrounding and related to Bell's inequalities and the foundations of quantum mechanics that must also be dealt with, many of an endlessly confounding and mysterious nature. Concerning the partial application of the noncommutative trigonometry to quantum computing I outlined in [46], more work needs to be done.

M. Seddighin has recently returned to the Re and Total antieigenvalue theory and has advanced our results of [16], [18] in his papers [75], [76], [77]. Our recent joint paper [59] grew out of those investigations. However it remains my feeling that one needs a better theory for arbitrary matrices than that provided by my original 1969 formulation of Real, Imaginary, and Total antieigenvalues.

When working on the optoelectronic neural networks, in 1992 my Ph.D. student Shelly Goggin and I implemented an angular version of the famous machine learning algorithm called Backpropagation. We called our algorithm Angleprop and a brief summary of the simulations may be found in [78].

The $C(H)$ result (4.19) referred to in the previous section is in [79]. I will come back to this in the next section.

Generally I cannot cite full literature here, which must be obtained from each paper and each area of application. One of the original treatments and still a good book for the ADI and other iterative algorithms is that of Varga [80]. However, the computational linear algebra literature has exploded in the last thirty years due to the availability of large scale computing. Of course the statistics literature and the quantum mechanics literature are also enormous.

Our interest in wavelets actually began with the stochastic processes paper [81], from which we developed a theory of Time operators. From our formulation of a Time operator for Kolmogorov Systems, I. Antoniou and I saw in 1991 that wavelets were wandering vectors: this discovery is documented in our NATO proposal [82]. I mention this here because at about the same time, others, e.g., see [83], also noticed this fact. For us it was natural from the beginning, because we were at that time working with the Sz. Nagy–Foias dilation theory [84], where wandering vectors occur on page 1. However, due to a number of more immediate tasks, we did not get around to publishing our results on wavelets until the papers [85], [86], [87], [88], although I did mention our results in the earlier books [23], [29].

In particular, we showed that every wavelet multisolution analysis (MRA) has naturally associated to it a Time operator T whose age eigenstates are the wavelet detail subspaces W_n. The natural setting of the Haar's wavelet is that of an eigenbasis canonically conjugate to the Haar detail refinement. The Time operator of the wavelet may be regarded as a (position) multiplication operator, equivalent under appropriate transform to a first order (momentum) differential operator.

Commentary. As concerns the resistance I sometimes encounter when bringing my operator trigonometry into a new domain of endeavor, I remember a discussion I once had with the late and renowned physicist (e.g., see [89]) Joseph Jauch in Switzerland in 1972. He mused that the influential mathematician Paul Halmos possessed and practiced a clever tactic: apply the same key insight or philosophical viewpoint to a variety of situations. As a result, no one could really beat Halmos at his game. There may be some parallel in the way in which the operator trigonometry has been developed.

6. Noncommutative trigonometry: Outlook

After I returned to this noncommutative operator trigonometry about a dozen years ago, the circumstances more or less described in Section 4, it became clear to me that it possessed the joint potentials for both further application and further theoretical development. During these last dozen years in which I have been able to develop these joint potentials, it has become clear to me that I cannot do it all myself. Toward the future enlargement of this theory and its applications, in the following three subsections I will indicate three fruitful future directions: Operator Algebras, Multiscale System Theory, and Quantum Mechanics.

6.1. Extensions to matrix and operator algebras

First, I would like to prove the result [90] announced in [54], and described in the commentary at the end of Section 4, heretofore unpublished, that an antieigenmatrix can be found which attains the maximum turning which we now understand (4.18) to mean. The notion of antieigenmatrix was first, to my knowledge, presented in my 1999 preprint [41]. However, it must be said the statisticians had already implicitly for many years had such antieigenmatrices lurking within their inequalities, although they did not see the geometry, hence they could not see the concept. Perhaps elsewhere I will try to elaborate all the places where such antieigenmatrices were already implicitly hiding in statistics.

In any case, to construct an antieigenmatrix for (4.18), I will follow here very closely and use the analysis of [79], where (4.19) was shown. The reader will need to consult [79] carefully to check all entities that I will use now.

There are two key steps in the proof in [79]. First it is shown that $C(H) = ||H'(A)||_F$ where $H'(A)$ is the Fréchet derivative. From this it follows by use of the implicit function theorem that in a neighborhood of A one has

$$\begin{aligned} H'(A)\Delta A &= \Delta H = VKV^T \\ &= V(\Gamma_2 \circ B + \Gamma_2^T \circ B^T)V^T \end{aligned} \tag{6.1}$$

where $\Gamma_2 = [\sigma_i/(\sigma_i + \sigma_j)]$, $B = U\Delta AV$, A has singular value decomposition $A = U\Sigma V^T$, $H = V\Sigma V^T$, and where $\circ$ denotes Hadamard product. The second key step, [79] Appendix, is to show that from considerations involving the structure of Γ_2 and Γ_2^T, one has

$$\begin{aligned} C(H) = ||H'(A)||_F &= \max_{||B||_F=1} ||\Gamma_2 \circ B + \Gamma_2^T \circ B^T||_F \\ &= \sqrt{2}\max_{i<j} \frac{\sqrt{1+(\sigma_i/\sigma_j)^2}}{1+\sigma_i/\sigma_j} \end{aligned} \tag{6.2}$$

from which the condition number bound (4.19) follows.

We may now connect to the dynamics of the operator trigonometry by seeking a ΔA which represents a maximal turning in the sense of the operator trigonometry and which attains the absolute condition number $C(H)$ of A. In other words, we seek an 'antieigenmatrix'. The Frobenius norm being unitarily invariant and recalling that $\Delta A = U^T BV^T$, it is easier to work with B as seen from (6.1) above. The following shows that $C(H)$ is exactly caused by such a sought maximal turning antieigenmatrix.

Proposition 6.1. *The absolute condition number $C(H)$ of the absolute factor H of the matrix $A = QH$ is attained by*

$$B = \begin{bmatrix} 0 & \cdots & 0 & \frac{\sigma_1}{(\sigma_1^2+\sigma_n^2)^{1/2}} \\ \vdots & & & \vdots \\ 0 & & & 0 \\ \frac{\sigma_n}{(\sigma_1^2+\sigma_n^2)^{1/2}} & & 0 \cdots & 0 \end{bmatrix}. \tag{6.3}$$

Proof. Following [79] and from (4.18) and (6.1), one has

$$C(H) = ||H'(A)||_F = \sup_{||\Delta A||_F=1} ||H'(A)\Delta A||_F = \sup_{||B||_F=1} ||2Re(\Gamma_2 \circ B)||_F. \tag{6.4}$$

However, rather than use the Hadamard product $\Gamma_2 \circ B$ formulation, we prefer to directly maximize $||K||_F$ from (6.1). Recall that $K = V^T \Delta H V = [k_{ij}]$ where $k_{ij} = (\sigma_i b_{ij} + \sigma_j b_{ji})/(\sigma_i + \sigma_j)$. The way we wish to think is the following: ΔA running through all directions is the same as B running through all directions from which we may look at all resulting symmetric K within which the data (the given matrix A's singular values) is already incorporated. First suppose B is symmetric. Then $k_{ij} = b_{ij}$ and $||\Delta H||_F = ||K||_F = ||B||_F = 1$ so such B are not interesting. A little thought and experimentation then led to the candidate B of (6.3). Consider first the 2×2 case and the matrix

$$K = \begin{bmatrix} b_{11} & \frac{\sigma_1 b_{12}+\sigma_2 b_{21}}{\sigma_1+\sigma_2} \\ \frac{\sigma_2 b_{21}+\sigma_1 b_{12}}{\sigma_2+\sigma_1} & b_{22} \end{bmatrix}. \tag{6.5}$$

We wish to maximize the expression

$$b_{11}^2 + b_{22}^2 + \frac{2(\sigma_1 b_{12} + \sigma_2 b_{21})^2}{(\sigma_1 + \sigma_2)^2} \tag{6.6}$$

subject to the constraint

$$b_{11}^2 + b_{22}^2 + b_{12}^2 + b_{21}^2 = 1. \tag{6.7}$$

We may take $b_{11}^2 = b_{22}^2 = 0$ and letting $x = b_{12}$ and $y = b_{21}$ we may maximize the expression $f(x) = \sigma_1 x + \sigma_2(1 - x^2)^{1/2}$, from which

$$x = \frac{\kappa}{(1+\kappa^2)^{1/2}}, \quad y = \frac{1}{(1+\kappa^2)^{1/2}} \tag{6.8}$$

where $\kappa = \sigma_1/\sigma_2$ ($K(A)$ in the notation of [79]) is the condition number of A. From (6.6) and (6.8) we have

$$||K||_F^2 = \frac{2}{(\sigma_1 + \sigma_2)^2} \frac{(\sigma_1 \kappa + \sigma_2)^2}{1 + \kappa^2} = \frac{2(\sigma_1^2 + \sigma_2^2)}{(\sigma_1 + \sigma_2)^2} = C^2(H). \tag{6.9}$$

The $n \times n$ situation is the same (choose all $b_{ij} = 0$ except b_{1n} and b_{n1} as given in (6.3)). Thus B of (6.3) enables K hence ΔA to attain $C(H)$. □

We may obtain a deeper geometrical clarification as follows.

Corollary 6.2. *With the B of (6.3), the matrix K has turning angle $\phi(H^2)$ on the antieigenvectors $x_\pm$ of H^2.*

Proof. We recall (2.8) that the (normalized) antieigenvectors of H^2, i.e., those vectors most turned by H^2, are

$$\begin{aligned} x_\pm &= \pm\frac{\sigma_1}{(\sigma_1^2+\sigma_n^2)^{1/2}}x_n + \frac{\sigma_n}{(\sigma_1^2+\sigma_n^2)^{1/2}}x_1 \\ &= \frac{1}{(\sigma_1^2+\sigma_n^2)^{1/2}}(\pm\sigma_1, 0, \ldots, 0, \sigma_n)^T \end{aligned} \tag{6.10}$$

where x_1 and x_n are the eigenvectors corresponding to σ_1 and σ_n, respectively.

Noting that K is the symmetric matrix with all $k_{ij} = 0$ except for

$$k_{1n} = k_{n1} = \frac{\sigma_1\kappa + \sigma_n}{(1+\kappa^2)^{1/2}(\sigma_1+\sigma_n)} = \frac{(\sigma_1^2+\sigma_n^2)^{1/2}}{(\sigma_1+\sigma_n)} \tag{6.11}$$

and since $||x_+|| = 1$, we find

$$\frac{\langle Kx_+, x_+\rangle}{||Kx_+||||x_+||} = \frac{2\sigma_1\sigma_n(\sigma_1^2+\sigma_n^2)^{-1/2}(\sigma_1+\sigma_n)^{-1}}{(\sigma_n^2+\sigma_1^2)^{1/2}(\sigma_1+\sigma_n)^{-1}} = \frac{2\sigma_1\sigma_n}{\sigma_1^2+\sigma_n^2} \tag{6.12}$$

which is (see (2.7)) exactly $\cos\phi(H^2)$. For x_- it is the same. □

Thus K, although not a positive definite matrix, nonetheless turns the antieigenvectors $x_\pm$ of $H^2 = A^*A$ in the same (maximal) way that H^2 does, and this is what causes B, which is constructed from H^2 antieigenvectors, to attain $C(H)$.

A more complete presentation [90] of this attainment of $C(H)$ by antieigenmatrices, and related considerations, will hopefully be published elsewhere. A natural remaining question is: characterize all of the antieigenmatrices which achieve the maximum turning angle.

But beyond presenting these results here which are new, we see that we have illustrated that the noncommutative operator trigonometry can be extended from geometrical action on vectors to geometrical action on matrices. In other words, in the future, hopefully to algebras of operators. If possible, it will be desirable to keep Hilbert space structures in the formulation, for trigonometry to be most natural needs a Euclidean geometry. The use of the Frobenius norm does do this since it has an inner product $tr(AB^*)$. On the other hand, the Frobenius norm is not an induced norm and $||I||_F = \sqrt{n}$ means that the key entity (2.4) will have to be modified. These are preliminary, tentative thoughts only, as also will be the following.

Much of C^* and W^* algebra tries to use commutativity when possible, if not, then commutators, if not, then commutants, or centers. But even when matrices or operators do not commute, one has the often times very useful fact that the spectra $\sigma(AB)$ and $\sigma(BA)$ are the same (for invertible operators, and differing only by ± 0 in the general case). This fact should somehow be regarded as a key invariant in future noncommutative trigonometries. This is a vague statement. But for example, when A and B are positive selfadjoint operators, and do not commute, then BA is not even normal, yet $\sigma(BA)$ is real and just from this spectrum one can define some parts of a (noncommutative) operator trigonometry. Then (e.g., see

[55]) one can find a whole family of positive selfadjoint operators C for which BA becomes positive selfadjoint in the C inner product. In this latter situation, we have a very comfortable, intuitive operator trigonometry, resembling the original one.

6.2. Multiscale system theory, wavelets, iterative methods

As is well known and as I mentioned in Section 4, there are strong links between wavelet methods, multilevel numerical methods, and multigrid and multiscale computing. I would like to elaborate that theme here. In particular, I want to advance the notion that Iterative methods should be added to this list, or at least placed into strong conjunction with these topics.

As I described in Section 4, at the Jackson, Wyoming Iterative Methods conference in 1997 (reported in [31]), I showed that the wavelet reconstruction algorithm, which had been developed independently by the wavelet community, was just the classical Richardson Iterative method. That method goes back to 1910 [91]. Now, in addition we know [31] the optimal wavelet reconstruction algorithm convergence rate to be operator trigonometric: $\sin\phi(S)$, where S is the frame operator. So wavelet reconstruction is both iterative and trigonometric.

Domain decomposition methods and multigrid numerical methods are clearly important multiscale methods. However, until the important analysis of Griebel and Oswald [92], it was not known that domain decomposition and multigrid methods could be placed within the context of Jacobi iterative methods. Because my operator trigonometric analysis applies to Jacobi methods, I could then place domain decomposition and multigrid methods into the noncommutative trigonometry. For more details see [30, 50].

The dramatic convergence improvements of the ADI semi-iterative methods described in Section 4 are clearly multiscale. So therefore are the underlying intrinsic noncommutative trigonometric subangles [36]. I should mention that to my knowledge, there is not yet a full theoretical understanding of the very important ADI schemes and their later derivatives. You can see the difficulties from the excellent early analysis of Varga [80]. You can see that only a commutative situation is easily analyzed. There is a very interesting noncommutative trigonometry of ADI waiting to be understood. The same can be said for the important SOR method combined with iterative refinement on a sequence of grids, see the preliminary discussion and results in [50].

For the PDE Dirichlet–Poisson model problem, a grid noncommutative trigonometry was worked out in [36]. The discretized operator A_h had turning angle $\phi(A_h) = \pi/2 - \pi h$, where h is the grid size. One will find in [36] also a discussion of how an operator mesh trigonometry shows that the performance of an iterative method on a grid of mesh size h depends upon the operator trigonometry on the finer mesh size $h/2$. This phenomenon is an aspect of interesting general identities within the operator trigonometry. For example [27],

$$\sin\phi(A^{1/2}) = \frac{\sin\phi(A)}{1+\cos\phi(A)} \tag{6.13}$$

for symmetric positive definite matrices A. Such power-of-two refinement relations are reminiscent of those of the dilations in wavelet multiresolutions. Another such scale relation from [90] is inherent in the absolute condition number $C(H)$ discussed in Section 6.1 above:

$$C^2(H) = \frac{2\sin\phi(H)}{\sin\phi(H^2)}. \tag{6.14}$$

To date I have only cared about such identities as they relate to specific theoretical or numerical applications.

A general theme could be stated philosophically as follows. All multiscale, multigrid, wavelet theories and behaviors are inherently spectral. So is my operator trigonometry of antieigenvalues and antieigenvectors. Eigenvalues take you from largest to smallest scale in your model. Instead of a vibration or numerical grid based range of scales, which correspond to an eigenvalue spectral decomposition, the antieigenvalues and antieigenvectors correspond to physical or numerical turning angle scales.

Turning to another subject, our theory of wavelets as stochastic processes [23, 29, 82, 85, 86, 87, 88] is inherently multiscale, but in quite a different sense. The stochastic processes and Kolmogorov dynamical systems may be seen as based upon a sequence of increasing σ-algebras of increasingly rich underlying event spaces. These are sometimes called filtrations in the stochastic literature, e.g., see [93], [94]. This type of stochastic event-multiscale is related to coarse-graining in statistical physics, and to conditional expectations in general. The Time operator of wavelets which we obtained [82,88], although motivated by mixing processes, is very much spectral. See my analysis in the recent paper [95], where I take it all back to a cyclic vector.

Our approach to wavelets in the above mentioned papers, although originally motivated by our work in statistical mechanics, can also be described as that of "representation-free theory of shifts," e.g., the Bernoulli shifts of dynamical systems theory being just one representation. In retrospect and as now connected to wavelet multiresolutions, we dealt first with the shift V (the dilation in wavelets) via its wandering subspaces W_n, before dealing with the translations. This enabled our getting the Time operator of wavelets. The conventional wavelet theory does not do that. Following our approach, Kubrusly and Levan [96], [97] showed that with respect to an orthonormal wavelet $\psi \in \mathcal{L}^2(\mathbb{R})$, any $f \in \mathcal{L}^2(\mathbb{R})$ is the sum of its layers of detail over all time shifts. This contrasts with the conventional wavelet view of f as the sum of its layers over all scales. Thus one may write

$$f(x) = \sum_m \sum_n \langle f, \psi_{m,n}\rangle \psi_{m,n}(x) = \sum_n \sum_m \langle f, \psi_{m,n}\rangle \psi_{m,n}. \tag{6.15}$$

I could describe this result as a kind of Fubini theorem, holding in the absence of the usual assumptions of absolute summability. More such results are to be expected from our point of view [87] of wavelets.

6.3. Quantum mechanics

Much of the current interest (and funding) for research on multiscale theory is the need in engineering applications to bring quantum mechanical effects into the modelling which has heretofore been only classical. I know this from many annecdotal conversations with my engineering colleagues. The applications range from chip design to aerodynamics. Each of those applications is a subject unto itself and will not be dealt with here. The considerations and needs are usually both physical and numerical.

Instead, I want to address quantum mechanics at a more fundamental level. As I mentioned briefly in Section 4, in [37, 42, 44, 46, 47, 49, 51, 52, 53, 60] I have found fundamental connections of the noncommutative trigonometry to quantum mechanics. There is too much in those papers to fully describe here, but let me at least delineate the issues and papers for bibliographical sake. Then I want to announce a new result here.

In [37, 42, 44, 47, 51, 52, 53] I was able to place the celebrated Bell inequality theory into the context of the noncommutative trigonometry. This was a completely new contribution to the foundations of physics. In particular, from my viewpoint, one cannot argue 'nonlocality' on the basis of violation of Bell's inequalities. See the papers for more details.

In [46] I develop preliminary aspects of the noncommutative trigonometry for quantum computing. To treat the unitary operators one encounters there, it becomes clear that I need a better complex-valued noncommutative operator trigonometry. That is to say, the general operator trigonometry which I developed in [45] for arbitrary operators A, by writing them in polar form $A = U|A|$ and then defining an operator trigonometry for them in terms of the known operator trigonometry for $|A|$, is not sufficient. Granted, quantum mechanics fundamentals often ignore phase when dealing just with probabilities. But for quantum computing, one cannot ignore phase.

In [49] I connect the noncommutative trigonometry to elementary particle physics. In particular, I showed that CP violation may be seen as a certain slightly unequal weighting of the CP antieigenvectors. The setting is that of Kaon systems and the decay of neutral mesons into charged pions. In particular, I showed that CP eigenstates are strangeness total antieigenvectors. CP violation is seen as antieigenvector-breaking.

Finally, I wish to turn to the famous quantum Zeno's paradox [98]. I was involved in its initial formulations in 1974 and a summary of the state of affairs may be found in [60], see also [99]. Mathematically and briefly, one is confronted with questions of the existence and nature of operator limits such as

$$\text{s-}\lim_{n\to\infty} (Ee^{itH}E)^n \tag{6.16}$$

where (roughly) E is (for example) a projection operator corresponding to a physical measuring apparatus, and H is a quantum mechanical Hamiltonian governing a wave packet evolution. When E is in the spectral family of H, i.e., when E

and H commute, all the issues are trivial, or otherwise may be dealt with in straightforward fashion. But in the physically interesting case when E and H do not commute, very interesting mathematical unbounded operator domain questions arise. Several enlightened physicists pushed me to return to these issues. A clarification, essentially known to me in 1974 but not published, is the following [60,99].

Lemma 6.3. *Let H be an unbounded selfadjoint operator and let E be an orthogonal projection. Then the domain $\mathcal{D}(HE)$ is dense iff EH is closeable. In that case,*

$$(HE)^* = \overline{EH} \supset EH \tag{6.17}$$

is defined, with domain at least as large as $\mathcal{D}(H)$. Generally $(EHE)^* = \overline{EH}E$ whenever EH is closeable. Furthermore the polar factors satisfy

$$|\overline{EH}| = (HE^2H)^{1/2}, \quad |HE| \supset (EH^2E)^{1/2} \tag{6.18}$$

Here HE^2H denotes $(\overline{EH})^*EH$.

Most of the recent papers by physicists on the Zeno issues, theory, and applications, fail to understand these unbounded noncommutative operator considerations. The critical rigor is not seen or at best is just glossed over or ignored. Another approach, that of Misra and Sudarshan originally [98], was to go to the Heisenberg picture and only work with the density matrix (bounded operators). Even there one eventually must assume operator limits such as (6.16).

Recently, in pondering some of these Bell, Zeno, or other related issues in the foundations of quantum mechanics, I found [100] the following operator theoretic characterization of quantum mechanical reversibility in terms of what I shall call domain regularity preservation. I have never seen this theorem elsewhere. I believe it to be fundamental to the understanding of quantum probability in the Schrödinger picture. I include it here because I see it as truly multiscale.

Theorem 6.4. *A unitary group U_t necessarily exactly preserves its infinitesimal generator's domain $\mathcal{D}(H)$. That is, U_t maps $\mathcal{D}(H)$ one to one onto itself, at every instant $-\infty < t < \infty$.*

The proof of Theorem 6.4 is not hard (I give two proofs in [100]) and the result is quite evident, once one sees it. Its application to quantum mechanics is dramatic: a quantum mechanical evolution must at all times be able to simultaneously account for each wave function probability upon which the Hamiltonian can act before one is entitled to draw any conclusions about the overall evolving probabilities. Not one probability can be lost.

A converse is the following [100]. Let Z_t be a contraction semigroup on a Hilbert space $\mathcal{H}$ such that Z_t maps $\mathcal{D}(A)$ one to one onto $\mathcal{D}(A)$. Here A is the infinitesimal generator of the semigroup Z_t. Then Z_t on the graph Hilbert space $\mathcal{H}_A$ can be extended to a group $Z_{-t} = Z_t^{-1}$.

What Theorem 6.4 means is that in unitary quantum mechanical evolutions, one cannot lose any of the totality of detail embodied in the totality of wave functions ψ in $\mathcal{D}(H)$. It is quite interesting to think about how much 'mixing around' of the $\mathcal{D}(H)$ probabilities the evolution can do. That is a multiscale and even interscale concept.

References

[1] K. Gustafson, *A Perturbation Lemma*, Bull. Amer. Math. Soc. **72** (1966), 334–338.

[2] K. Gustafson, *Positive Operator Products*, Notices Amer. Math. Soc. **14** (1967), Abstract 67T-531, p. 717. See also Abstracts 67T-340, 67T-564, 67T-675.

[3] K. Gustafson, *A Note on Left Multiplication of Semi-group Generators*, Pacific J. Math. **24** (1968a), 463–465.

[4] K. Gustafson, *The Angle of an Operator and Positive Operator Products*, Bull. Amer. Math. Soc. **74** (1968b), 488–492.

[5] K. Gustafson, *Positive (noncommuting) Operator Products and Semigroups*, Math. Zeitschrift **105** (1968c), 160–172.

[6] K. Gustafson, *A min–max Theorem*, Amer. Math Soc. Notices **15** (1968d), p. 799.

[7] K. Gustafson, *Doubling Perturbation Sizes and Preservation of Operator Indices in Normed Linear Spaces*, Proc. Camb. Phil. Soc. **98** (1969a), 281–294.

[8] K. Gustafson, *On the Cosine of Unbounded Operators*, Acta Sci. Math. **30** (1969b), 33–34 (with B. Zwahlen).

[9] K. Gustafson, *Some Perturbation Theorems for Nonnegative Contraction Semigroups*, J. Math. Soc. Japan **21** (1969c), 200–204 (with Ken-iti Sato).

[10] K. Gustafson, *A Simple Proof of the Toeplitz–Hausdorff Theorem for Linear Operators*, Proc. Amer. Math. Soc. **25** (1970), 203–204.

[11] K. Gustafson, *Anti-eigenvalue Inequalities in Operator Theory*, Inequalities III (O. Shisha, ed.), Academic Press (1972a), 115–119.

[12] K. Gustafson, *Multiplicative Perturbation of Semigroup Generators*, Pac. J. Math. **41** (1972b), 731–742 (with G. Lumer).

[13] K. Gustafson, *Multiplicative Perturbation of Nonlinear m-accretive Operators*, J. Funct. Anal. **10** (1972c), 149–158 (with B. Calvert).

[14] K. Gustafson, *Numerical Range and Accretivity of Operator Products*, J. Math. Anal. Applic. **60** (1977), 693–702 (with D. Rao).

[15] K. Gustafson, *The RKNG (Rellich, Kato, Nagy, Gustafson) Perturbation Theorem for Linear Operators in Hilbert and Banach Space*, Acta Sci. Math. **45** (1983), 201–211.

[16] K. Gustafson, *Antieigenvalue Bounds*, J. Math. Anal. Applic. **143** (1989), 327–340 (with M. Seddighin).

[17] K. Gustafson, *Antieigenvalues in Analysis*, Fourth International Workshop in Analysis and its Applications, Dubrovnik, Yugoslavia, June 1–10, 1990 (C. Stanojevic and O. Hadzic, eds.), Novi Sad, Yugoslavi (1991), 57–69.

[18] K. Gustafson, *A Note on Total Antieigenvectors*, J. Math. Anal. Applic. **178** (1993), 603–611 (with M. Seddighin).

[19] K. Gustafson, *Operator Trigonometry*, Linear and Multilinear Algebra **37** (1994a), 139–159.

[20] K. Gustafson, *Antieigenvalues*, Lin. Alg. and Applic. **208/209** (1994b), 437–454.

[21] K. Gustafson, *Computational Trigonometry*, Proc. Colorado Conf. on Iterative Methods, Vol. 1 (1994c), p. 1.

[22] K. Gustafson, *Matrix Trigonometry*, Lin. Alg. and Applic. **217** (1995), 117–140.

[23] K. Gustafson, *Lectures on Computational Fluid Dynamics, Mathematical Physics, and Linear Algebra*, Kaigai Publishers, Tokyo, Japan (1996a), 169 pp.

[24] K. Gustafson, *Trigonometric Interpretation of Iterative Methods*, Proc. Conf. Algebraic Multilevel Iteration Methods with Applications, (O. Axelsson, B. Polman, eds.), Nijmegen, Netherlands, June 13–15, (1996b), 23–29.

[25] K. Gustafson, *Commentary on Topics in the Analytic Theory of Matrices, Section 23, Singular Angles of a Square Matrix*, Collected Works of Helmut Wielandt 2 (B. Huppert and H. Schneider, eds.), De Gruyters, Berlin (1996c), 356–367.

[26] K. Gustafson, *Numerical Range: The Field of Values of Linear Operators and Matrices*, Springer, Berlin, (1997a), pp. 205 (with D. Rao).

[27] K. Gustafson, *Operator Trigonometry of Iterative Methods*, Num. Lin Alg. with Applic. **34** (1997b), 333–347.

[28] K. Gustafson, *Antieigenvalues*, Encyclopaedia of Mathematics, Supplement 1, Kluwer Acad. Publ., Dordrecht, (1997c), 57.

[29] K. Gustafson, *Lectures on Computational Fluid Dynamics, Mathematical Physics, and Linear Algebra*, World Scientific, Singapore, (1997d), pp. 178.

[30] K. Gustafson, *Domain Decomposition, Operator Trigonometry, Robin Condition*, Contemporary Mathematics **218** (1998a), 455–560.

[31] K. Gustafson, *Operator Trigonometry of Wavelet Frames*, Iterative Methods in Scientific Computation (1998b), 161–166; (J. Wang, M. Allen, B. Chen, T. Mathew, eds.), IMACS Series in Computational and Applied Mathematics **4**, New Brunswick, NJ.

[32] K. Gustafson, *Semigroups and Antieigenvalues*, Irreversibility and Causality–Semigroups and Rigged Hilbert Spaces, (A. Bohm, H. Doebner, P. Kielanowski, eds.), Lecture Notes in Physics **504**, Springer, Berlin, (1998c), pp. 379–384.

[33] K. Gustafson, *Operator Trigonometry of Linear Systems*, Proc. 8th IFAC Symposium on Large Scale Systems: Theory and Applications, (N. Koussoulas, P Groumpos, eds.), Patras, Greece, July 15–17, (1998d), 950–955. (Also published by Pergamon Press, 1999.)

[34] K. Gustafson, *Symmetrized Product Definiteness? Comments on Solutions 19-5.1–19-5.5*, IMAGE: Bulletin of the International Linear Algebra Society **21** (1998e), 22.

[35] K. Gustafson, *Antieigenvalues: An Extended Spectral Theory*, Generalized Functions, Operator Theory and Dynamical Systems, (I. Antoniou, G. Lumer, eds.), Pitman Research Notes in Mathematics **399**, (1998f), 144–149, London.

[36] K. Gustafson, *Operator Trigonometry of the Model Problem*, Num. Lin. Alg. with Applic. **5** (1998g), 377–399.

[37] K. Gustafson, *The Geometry of Quantum Probabilities*, On Quanta, Mind, and Matter: Hans Primas in Context, (H. Atmanspacher, A. Amann, U. Mueller–Herold, eds.), Kluwer, Dordrecht (1999a), 151–164.

[38] K. Gustafson, *The Geometrical Meaning of the Kantorovich–Wielandt Inequalities*, Lin. Alg. and Applic. **296** (1999b), 143–151.

[39] K. Gustafson, *Symmetrized Product Definiteness: A Further Comment*, IMAGE: Bulletin of the International Linear Algebra Society **22** (1999c), 26.

[40] K. Gustafson, *A Computational Trigonometry and Related Contributions by Russians Kantorovich, Krein, Kaporin*, Computational Technologies **4** (No. 3) (1999d), 73–83, (Novosibirsk, Russia).

[41] K. Gustafson, *On Geometry of Statistical Efficiency*, (1999e), preprint.

[42] K. Gustafson, *The Trigonometry of Quantum Probabilities*, Trends in Contemporary Infinite-Dimensional Analysis and Quantum Probability, (L. Accardi, H. Kuo, N. Obata, K. Saito, S. Si, L. Streit, eds.), Italian Institute of Culture, Kyoto (2000a), 159–173.

[43] K. Gustafson, *Semigroup Theory and Operator Trigonometry*, Semigroups of Operators: Theory and Applications, (A.V. Balakrisnan, ed.), Birkhäuser, Basel (2000b), 131–140.

[44] K. Gustafson, *Quantum Trigonometry*, Infinite-Dimensional Analysis, Quantum Probability, and Related Topics **3** (2000c), 33–52.

[45] K. Gustafson, *An Extended Operator Trigonometry*, Lin. Alg. & Applic. **319** (2000d), 117–135.

[46] K. Gustafson, *An Unconventional Linear Algebra: Operator Trigonometry*, Unconventional Models of Computation, UMC'2K, (I. Antoniou, C. Calude, M. Dinneen, eds.), Springer, London (2001a), 48–67.

[47] K. Gustafson, *Probability, Geometry, and Irreversibility in Quantum Mechanics*, Chaos, Solitons and Fractals **12** (2001b), 2849–2858.

[48] K. Gustafson, *Operator Trigonometry of Statistics and Econometrics*, Lin. Alg. and Applic. **354** (2002a), 151–158.

[49] K. Gustafson, *CP-Violation as Antieigenvector-Breaking*, Advances in Chemical Physics **122** (2002b), 239–258.

[50] K. Gustafson, *Operator Trigonometry of Preconditioning, Domain Decomposition, Sparse Approximate Inverses, Successive Overrelaxation, Minimum Residual Schemes* Num. Lin Alg. with Applic. **10** (2003a), 291–315.

[51] K. Gustafson, *Bell's Inequalities*, Contributions to the XXII Solvay Conference on Physics, (A. Borisov, ed.), Moscow–Izhevsk, ISC, Moscow State University (ISBN: 5-93972-277-6) (2003b), 501–517.

[52] K. Gustafson, *Bell's Inequality and the Accardi–Gustafson Inequality*, Foundations of Probability and Physics-2, (A. Khrennikov, ed.) Växjo University Press, Sweden (2003c), 207–223.

[53] K. Gustafson, *Bell's Inequalities*, The Physics of Communication, Proceedings of the XXII Solvay Conference on Physics, (I. Antoniou, V. Sadovnichy, H. Walther, eds.), World Scientific (2003d), 534–554.

[54] K. Gustafson, *Preconditioning, Inner Products, Normal Degree*, 2003 International Conference on Preconditioning Techniques for Large Sparse Matrix Problems in Scientific and Industrial Applications, (E. Ng, Y. Saad, W.P. Tang, eds.), Napa, CA, 27–29, October (2003e), pp. 3.

[55] K. Gustafson, *An Inner Product Lemma*, Num. Lin. Alg. with Applic. **11** (2004a), 649–659.

[56] K. Gustafson, *Normal Degree*, Num. Lin. Alg. with Applic. **11** (2004b), 661–674.

[57] K. Gustafson, *Interaction Antieigenvalues*, J. Math Anal. Applic. **299** (2004c), 174–185.

[58] K. Gustafson, *The Geometry of Statistical Efficiency*, Research Letters Inf. Math. Sci. **8** (2005a), 105–121.

[59] K. Gustafson, *On the Eigenvalues Which Express Antieigenvalues*, International J. of Mathematics and Mathematical Sciences 2005:**10** (2005b), 1543–1554. (with M. Seddighin).

[60] K. Gustafson, *Bell and Zeno*, International J. of Theoretical Physics **44** (2005c), 1931–1940.

[61] M. Krein, *Angular Localization of the Spectrum of a Multiplicative Integral in a Hilbert Space*, Funct. Anal. Appl. **3** (1969), 89–90.

[62] H. Wielandt, *Topics in the Analytic Theory of Matrices*, University of Wisconsin Lecture Notes, Madison, (1967).

[63] E. Asplund and V. Ptak, *A Minmax Inequality for Operators and a Related Numerical Range*, Acta Math. **126** (1971), 53–62.

[64] P. Hess, *A Remark on the Cosine of Linear Operators*, Acta Sci Math. **32** (1971), 267–269.

[65] C. Davis, *Extending the Kantorovich Inequalities to Normal Matrices*, Lin. Alg. Appl. **31** (1980), 173–177.

[66] B.A. Mirman, *Antieigenvalues: Method of Estimation and Calculation*, Lin. Alg. Appl. **49** (1983), 247–255.

[67] L. Kantorovich, *Functional Analysis and Applied Mathematics*, Uspekhi Mat. Nauk **3** No. 6 (1948), 89–185.

[68] S. Drury, S. Liu, C.Y. Lu, S. Puntanen, and G.P.H. Styan, *Some Comments on Several Matrix Inequalities with Applications to Canonical Correlations: Historical Background and Recent Developments*, Sankhya: The Indian J. of Statistics **64**, Series A, Pt. 2 (2002), 453–507.

[69] R. Khatree, *On Calculation of Antiegenvalues and Antieigenvectors*, J. Interdisciplinary Math **4** (2001), 195–199.

[70] R. Khatree, *On Generalized Antieigenvalue and Antieigenmatrix of order r*, Amer. J. of Mathematical and Management Sciences **22** (2002), 89–98.

[71] R. Khatree, *Antieigenvalues and Antieigenvectors in Statistics*, J. of Statistical Planning and Inference **114** (2003), 131–144.

[72] K.C. Das, M. Das Gupta, K. Paul, *Structure of the Antieigenvectors of a Strictly Accretive Operator*, International J. of Mathematics and Mathematical Sciences **21** (1998), 761–766.

[73] C.R. Rao, *Anti-eigen and Anti-singular Values of a Matrix and Applications to Problems in Statistics*, Research Letters Inf. Math. Sci. 2005:**10** (2005), 53–76.

[74] K. Gustafson, R. Hartman, *Divergence-free Bases for Finite Element Schemes in Hydrodynamics*, SIAM J. Numer. Analysis **20** (1983), 697–721.

[75] M. Seddighin, *Antieigenvalues and Total Antieigenvalues of Normal Operators*, J. Math. Anal. Appl. **274** (2002), 239–254.

[76] M. Seddighin, *Antieigenvalue Inequalities in Operator Theory*, International J. of Mathematics and Mathematical Sciences (2004), 3037–3043.

[77] M. Seddighin, *On the Joint Antieigenvalues of Operators in Normal Subalgebras*, J. Math. Anal. Appl. (2005), to appear.

[78] K. Gustafson, *Distinguishing Discretization and Discrete Dynamics, with Application to Ecology, Machine Learning, and Atomic Physics*, in Structure and Dynamics of Nonlinear Wave Phenomena, (M. Tanaka, ed.), RIMS Kokyuroku 1271, Kyoto, Japan (2002c), 100–111.

[79] F. Chaitin–Chatelin, S. Gratton, *On the Condition Numbers Associated with the Polar Factorization of a Matrix*, Numerical Linear Algebra with Applications **7** (2000), 337–354.

[80] R. Varga, *Matrix Iterative Analysis*, Prentice Hall, NJ, (1962).

[81] K. Gustafson, B. Misra, *Canonical Commutation Relations of Quantum Mechanics and Stochastic Regularity*, Letters in Math. Phys. **1** (1976), 275–280.

[82] K. Gustafson, I. Antoniou, *Wavelets and Kolmogorov Systems*, http://www.auth.gr/chi/PROJECTSWaveletsKolmog.html, (2004).

[83] X. Dai, D.R. Larson, *Wandering Vectors for Unitary Systems and Orthogonal Wavelets*, Amer. Math. Soc. Memoirs **640**, Providence, RI, (1998).

[84] B. Sz. Nagy, C. Foias, *Harmonic Analysis of Operators in Hilbert Space*, North Holland, Amsterdam, (1970).

[85] I. Antoniou, K. Gustafson, *Haar Wavelets and Differential Equations*, Differential Equations **34** (1998), 829–832.

[86] K. Gustafson, *Wavelets as Stochastic Processes*, Workshop on Wavelets and Wavelet-based Technologies, (M. Kobayashi, S. Sakakibara, M. Yamada, eds.), IBM-Japan, Tokyo, October 29–30, (1998h), 40–43.

[87] I. Antoniou, K. Gustafson, *Wavelets and Stochastic Processes*, Mathematics and Computers in Simulation **49** (1999), 81–104.

[88] I. Antoniou, K. Gustafson, *The Time Operator of Wavelets*, Chaos, Solitons and Fractals **11** (2000), 443–452.

[89] J.M. Jauch, *Foundations of Quantum Mechanics*, Addison–Wesley, Reading, MA, (1968).

[90] K. Gustafson, *The Geometrical Meaning of the Absolute Condition Number of the Hermitian Polar Factor of a Matrix*, (2004d), preprint.

[91] L.F. Richardson, *The Approximate Arithmetical Solution by Finite Differences of Physical Problems Involving Differential Equations with an Application to the Stresses in a Masonry Dam*, Phil. Trans. Roy. Soc. London **A242** (1910), 307–357.

[92] M. Griebel, P. Oswald, *On the Abstract Theory of Additive and Multiplicative Schwarz Algorithms*, Numerische Mathematik **70** (1995), 163–180.

[93] P. Kopp, *Martingales and Stochastic Integrals*, Cambridge University Press, Cambridge (1984).

[94] R. Dudley, *Real Analysis and Probability*, Chapman & Hall, New York (1989).

[95] K. Gustafson, *Continued Fractions, Wavelet Time Operators, and Inverse Problems*, Rocky Mountain J. Math **33** (2003f), 661–668.

[96] N. Levan, C. Kubrusly, *A Wavelet "time-shift-detail" Decomposition*, Math. Comput. Simulation **63** (2003), 73–78.

[97] N. Levan, C. Kubrusly, *Time-Shifts Generalized Multiresolution Analysis over Dyadic-scaling Reducing Subspaces*, International J. Wavelets, Multiresolution and Information Processing **2** (2004), 237–248.

[98] B. Misra, G. Sudarshan, *The Zeno's Paradox in Quantum Theory*, J. of Mathematical Physics **18** (1977), 756–763.

[99] K. Gustafson, *The Quantum Zeno Paradox and the Counter Problem*, Foundations of Probability and Physics-2 (A. Khrennikov, ed.), Växjo University Press, Sweden (2003g), 225–236.

[100] K. Gustafson, *Reversibility and Regularity* (2004), preprint, to appear, International J. of Theoretical Physics (2006).

Karl Gustafson
Department of Mathematics
University of Colorado
Boulder, CO 80309-0395, USA
e-mail: gustafs@euclid.colorado.edu

Operator Theory:
Advances and Applications, Vol. 167, 157–171

Stationary Random Fields over Graphs and Related Structures

Herbert Heyer

Abstract. The concept of a stationary random field can be extended from the classical situation to a more general set up by letting the time parameter run through groups, homogeneous spaces or hypergroups. In the present exposition the author is concerned with spectral representations for stationary random fields over these algebraic-topological structures.

Mathematics Subject Classification (2000). Primary: 60G60. Secondary:43A62.

Keywords. generalized random fields, hypergroups.

1. Introduction

Stationary random fields over trees and buildings can be efficiently studied by looking at the algebraic-topological structure of such configurations. In their work [10] of 1989 R. Lasser and M. Leitner introduced the notion of stationary random fields over a commutative hypergroup, in particular over a double coset space $G//H$ arising from a Gelfand pair (G, H) consisting of a locally compact group G and a compact subgroup H of G. These authors put emphasis on some statistical problems, in particular on the estimation of the mean of stationary random sequences. Much earlier the hypergroup approach to stationary random functions over homogeneous spaces of the form G/H had been of significant interest to probabilists. A broad discussion of the theory in the case that G/H admits a Riemannian structure can be found in the paper [9] of R. Gangolli published in 1967. The first attempt to deal with stationary random functions over discrete homogeneous spaces G/H which appear as graphs or buildings goes back to J.P. Arnaud and G. Letac [3] who in 1984 gave a Karhunen representation in the case of a homogeneous tree. It is the aim of the present exposition to compare the approaches to stationary random fields over the discrete spaces $G//H$ and G/H and to pinpoint the technical differences in obtaining Karhunen-type representations.

In Chapter 1 we recapitulate the basics on second-order random fields leading to the classical Karhunen representation. Chapter 2 is devoted to stationary random fields over graphs and their corresponding polynomial hypergroups. Arnaud's Karhunen isometry contained in his work [2] of 1994 is considered to supplement section 8.2 of the monograph [4].

For the notation applied one observes that $\mathbb{Z}, \mathbb{R}$ denote the sets of integers and reals respectively. For a measurable space $(E, \mathfrak{E}), M(E) = M(E, \mathfrak{E})$ stands for the set of measures on $(E, \mathfrak{E})$, ε_x denoting the Dirac measure in $x \in E$. As usual $L^2_{\mathbb{K}}(E, \mathfrak{E}, \mu)$ symbolizes the space of μ-square integrable functions on $(E, \mathfrak{E})$ with values in $\mathbb{K} = \mathbb{C}$ or $\mathbb{R}$. Occasionally $\mathfrak{E}$ will be choosen to be the Borel-σ-algebra $\mathfrak{B}(E)$ of a topological space E. For such $E, C(E)$ is the space of continuous functions on E. The upper b and the lower $+$ attached to M and C refer to boundedness and positivity respectively.

2. Second-order random fields

2.1. Basic notions

A stochastic process $(\Omega, \mathfrak{A}, \mathbb{P}, \{X(t) : t \in T\})$ on a probability space $(\Omega, \mathfrak{A}, \mathbb{P})$ with parameter set T and state space $\mathbb{C}$ is called a (second-order) *random field over* T if $X(t) \in L^2_{\mathbb{C}}(\Omega, \mathfrak{A}, \mathbb{P})$ for every $t \in T$. In the sequel we shall view such a process as a mapping $t \mapsto X(t)$ from T into the space $L^2_{\mathbb{C}}(\Omega, \mathfrak{A}, \mathbb{P})$. Along with a random field X over T we define its *mean function*

$$t \mapsto m(t) := E(X(t))$$

and *covariance kernel*

$$\begin{aligned} (s,t) \mapsto C(s,t) : &= C(X(s), X(t)) \\ &= E[(X(s) - m(s))\overline{(X(t) - m(t))}] \end{aligned}$$

on T and $T \times T$ respectively. A random field X over T is said to be *centered* provided $m \equiv 0$.

Given a random field X over T one notes that its covariance kernel C is a *positive form* on $T \times T$ in the sense that

(a) $C(t,s) = \overline{C(s,t)}$ for all $(s,t) \in T$ and
(b) for each $n \geq 1, (t_1, \ldots, t_n) \in T^n$ and $(c_1, \ldots, c_n) \in \mathbb{C}^n$,

$$\sum_{i,j=1}^{n} c_i \overline{c_j}\, C(t_i, t_j) \geq 0.$$

It is one of the first significant facts in the theory of random fields that given functions $m : T \to \mathbb{R}$ and $C : T \times T \to \mathbb{R}$ such that for every $n \geq 1$ and every $(t_1, \ldots, t_n) \in T^n$ the matrix $(C(t_i, t_j))_{i,j=1,\ldots,n}$ is symmetric and positive, there exists a Gaussian process $(\Omega, \mathfrak{A}, \mathbb{P}, \{X(t) : t \in T\})$ whose finite-dimensional distribution $\mathbb{P}_{(X(t_1),\ldots,X(t_n))}$ is the normal distribution

$$N((m(t_i))_{i=1,\ldots,n}, (C(t_i, t_j))_{i,j=1,\ldots,n}) \qquad (\text{for } n \geq 1).$$

2.2. Spatial random fields with orthogonal increments

Let $(E, \mathfrak{E}, \mu)$ denote a σ-finite measure space. We consider the parameter set $T := \{A \in \mathfrak{E} : \mu(A) < \infty\}$. Then the random field Z over T is called a *spatial random field* (over T) with orthogonal *increments and base* μ provided the mapping $A \mapsto Z(A)$ satisfies the following conditions:

(a) $Z(\varnothing) = 0$.

(b) For $A, B \in T$ with $A \cap B = \varnothing$,

$$Z(A \cup B) = Z(A) + Z(B).$$

(c) For $A, B \in T$ we have that

$$C(Z(A), Z(B)) = \mu(A \cap B).$$

Z is said to be *centered* if $Z(A)$ is a centered random variable for all $A \in T$.

Examples of spatial random fields with orthogonal increments

(1) *Spatial Poisson random fields with intensity measure* μ.
They are defined for a given σ-finite measure space $(E, \mathfrak{E}, \mu)$ and a Markov kernel Z from a probability space $(\Omega, \mathfrak{A}, \mathbb{P})$ to $(E, \mathfrak{E})$ as processes

$$(\Omega, \mathfrak{A}, \mathbb{P}, \{Z(A) : A \in T\})$$

with parameter set $T = \{A \in \mathfrak{E} : \mu(A) < \infty\}$ and state space

$$\overline{\mathbb{Z}}_{+} \subset \overline{\mathbb{R}} := \mathbb{R} \cup \{-\infty, +\infty\}$$

having the following properties

(a) For each $n \geq 1$ and sequences $\{A_1, \ldots, A_n\}$ of pairwise disjoint sets in $T, \{Z(A_1), \ldots, Z(A_n)\}$ is a sequence of independent random variables.

(b) For each $A \in T$ the random variable $\omega \mapsto Z(\omega, A)$ admits the distribution

$$\mathbb{P}_{Z(A)} = \Pi(\mu(A)) = e^{-\mu(A)} \sum_{k \geq 0} \frac{\mu(A)^k}{k!} \varepsilon_k.$$

(2) *Spatial Gaussian random fields with independent increments and base* μ.
They are easily constructed by the method described in Section 2.1. In fact, given a σ-finite measure μ on the measurable space $(E, \mathfrak{E})$ one observes that the mapping

$$(A, B) \mapsto C(A, B) := \mu(A \cap B)$$

from $T \times T$ into $\mathbb{R}$ is symmetric and positive and hence that for any $n \geq 1$ and any sequence $\{A_1, \ldots, A_n\}$ of pairwise disjoint sets in T the matrix $(C(A_i, A_j))_{i,\,j=1,\ldots,\,n}$ is symmetric and positive. Then there exists a centered random field $(\Omega, \mathfrak{A}, \mathbb{P}, \{Z(A) : A \in T\})$ with parameter set $T := \{A \in \mathfrak{E} : \mu(A) < \infty\}$ and state space $\mathbb{R}$ such that

$$\mathbb{P}_{(Z(A_1),\ldots,\,Z(A_n))} = N(0, (C(A_i, A_j))_{i,\,j=1,\ldots,\,n}).$$

Subexamples.

(2.1) For the choices $E := \mathbb{R}^d, \mathfrak{C} := \mathfrak{B}(E)$ and $\mu :=$ d-dimensional Lebesgue measure one obtains the *Brownian field.*

(2.1.1) If $d = 1$, the process $(\Omega, \mathfrak{A}, \mathbb{P}, \{X(t) : t \in \mathbb{R}_+\})$ with $X(t) := Z([0,t])$ for all $t \in \mathbb{R}_+$ appears to be *Brownian motion* in $\mathbb{R}$.

Now, let Z be a spatial random field over T with orthogonal increments and base μ. For $f = \sum_{i=1}^{m} a_i 1_{A_i}$ and $g = \sum_{j=1}^{n} b_j 1_{B_j}$ in the space V of step functions in $L^2_{\mathbb{C}}(E, \mathfrak{C}, \mu)$ we obtain

$$\int f\bar{g}d\mu = C\left(\sum_{i=1}^{m} a_i\, Z(A_i), \sum_{j=1}^{n} b_j\, Z(B_j)\right).$$

Let

$$H^Z_{\mathbb{C}} := \langle\{Z(A) : A \in T\}\rangle^-$$

denote the Hilbert space generated by the random variables $Z(A) (A \in T)$. Clearly,

$$H^Z_{\mathbb{C}} \supset \left\{\sum_{i=1}^{m} a_i\, Z(A_i) : a_i \in \mathbb{C}, A_i \in T, i = 1,, \dots, m; m \geq 1\right\}.$$

From the fact that $V^- = L^2_{\mathbb{C}}(E, \mathfrak{C}, \mu)$ follows the subsequent

Theorem. There exists a unique isometry

$$L^2_{\mathbb{C}}(E, \mathfrak{C}, \mu) \to H^Z_{\mathbb{C}}$$

extending the mapping

$$1_A \mapsto Z(A) =: Z(1_A)$$

defined on the subset $\{1_A : A \in T\}$ of $L^2_{\mathbb{C}}(E, \mathfrak{C}, \mu)$.

For every $f \in L^2_{\mathbb{C}}(E, \mathfrak{C}, \mu)$

$$Z(f) = \int f\,dZ$$

is called the *stochastic integral* of f with respect to Z. Obviously, $Z(f)$ is centered for $f \in L^2_{\mathbb{C}}(E, \mathfrak{C}, \mu)$, and

$$C(Z(f), Z(g)) = \int f\bar{g}d\mu$$

whenever $f, g \in L^2_{\mathbb{C}}(E, \mathfrak{C}, \mu)$.

Examples of stochastic integrals.

(1) Let Z be a spatial Poisson random field with intensity μ. Then the random field $A \mapsto Z_1(A) := Z(A) - \mu$ is centered, and

$$Z_1(f) = Z(f) - \mu(f) = \int f dZ - \int f d\mu$$

for all $f \in L^2_{\mathbb{C}}(E, \mathfrak{C}, \mu)$.

(2) Let Z be a centered spatial Gaussian random field with base μ. Then

$$Z(f) \in H_{\mathbb{R}}^{Z},$$

hence $Z(f)$ is a Gaussian random variable with

$$\mathbb{P}_{Z(f)} = N(0, \mu(f))$$

for all $f \in L_{\mathbb{R}}^{2}(E, \mathfrak{E}, \mu)$.

2.3. The Karhunen representation

An important result on random fields is their stochastic integral representation with respect to a spatial random field with orthogonal increments. Our exposition follows that of D. Ducunha-Castelle and M. Duflo in their book [8] of 1986.

Theorem. Let T be an arbitrary set, and let $(E, \mathfrak{E}, \mu)$ be some measure space. For a given mapping $a : T \times E \to \mathbb{C}$ with $a(t, \cdot) \in L_{\mathbb{C}}^{2}(E, \mathfrak{E}, \mu)$ whenever $t \in T$ we define a mapping $C : T \times T \to \mathbb{C}$ by

$$C(s,t) := \int_E a(s,x)\overline{a(t,x)}\mu(dx)$$

for all $s, t \in T$.

Let X be a centered random field over T with covariance kernel C. Then there exists a spatial random field $A \mapsto Z(A)$ over $U := \{A \in \mathfrak{E} : \mu(A) < \infty\}$ with orthogonal increments and base μ such that

$$X(t) = \int_E a(t,x) Z(\cdot, dx)$$

for all $t \in T$. Moreover, the following statements are equivalent:

(i) $H_{\mathbb{C}}^{X} = H_{\mathbb{C}}^{Z}$.

(ii) $\langle \{a(t, \cdot) : t \in T\}\rangle^{-} = L_{\mathbb{C}}^{2}(E, \mathfrak{E}, \mu)$.

Remark. The statements of the theorem indicate the general task to be taken up once one wishes to establish the Karhunen isometry: At first one has to provide the mapping a satisfying condition (ii); secondly the spatial random field Z has to be constructed admitting

$$(s,t) \mapsto \int_E a(s,x)\, \overline{a(t,x)}\, \mu(dx)$$

as its covariance kernel.

The idea of the *proof* of the theorem admittedly well-known is worthwhile to be reproduced (from [8]).

$$E(| X(t) - X(s) |^2) = \int_E | a(t,x) - a(s,x) |^2 \mu(dx)$$

valid for all $s, t \in T$ implies that $a(t, \cdot) = a(s, \cdot)$ $\mu - a.s.$ if and only if $X(t) = X(s)$ $\mathbb{P} - a.s.$. Now let K denote the Hilbert space $\langle \{a(t, \cdot) : t \in T\}\rangle^{-}$ generated in $L_{\mathbb{C}}^{2}(E, \mathfrak{E}, \mu)$ by the functions $a(t, \cdot)$ $(t \in T)$.

The mapping

$$a(t, \cdot) \mapsto X(t)$$

has a unique extension to an isometry $\psi : K \to H_{\mathbb{C}}^X$.

Assume now that $K = L_{\mathbb{C}}^2(E, \mathfrak{E}, \mu)$. Then $1_A \in L_{\mathbb{R}}^2(E, \mathfrak{E}, \mu)$ for all $A \in U$. We put

$$Z(A) := \psi(1_A)$$

for all $A \in U$ and see that $A \mapsto Z(A)$ is a spatial random field over U with orthogonal increments and base μ. But then ψ is an isometry from $L_{\mathbb{R}}^2(E, \mathfrak{E}, \mu)$ onto $H_{\mathbb{C}}^Z$ taking 1_A into $Z(A)$ for all $A \in U$. From

$$X(t) = \psi(a(t, \cdot)) = \int_E a(t, x) Z(\cdot, dx) \ (t \in T)$$

we conclude that $H_{\mathbb{C}}^X \subset H_{\mathbb{C}}^Z$. But $Z(A) \in H_{\mathbb{C}}^X$, since $\mu(A) < \infty$ for all $A \in \mathfrak{E}$, hence $H_{\mathbb{C}}^X = H_{\mathbb{C}}^Z$.

In the case that $K \underset{+}{\subset} L_{\mathbb{C}}^2(E, \mathfrak{E}, \mu)$ we consider the orthogonal decomposition of K in $L_{\mathbb{C}}^2(E, \mathfrak{E}, \mu)$ with

$$K^{\perp} := \langle \{a(t', \cdot) : t' \in T', \} \rangle^{-},$$

where $T' \cap T = \varnothing$. We introduce

$$C(s, t) := \int_E a(s, x) \overline{a(t, x)} \, \mu(dx)$$

for all $s, t \in T \cup T'$, with $C(s, t) := 0$ whenever $s \in T, t \in T'$, a random field $t \mapsto X(t)$ as a mapping $T' \to L_{\mathbb{C}}^2(\Omega', \mathfrak{A}', \mathbb{P}')$ with covariance kernel C restricted to $T' \times T'$, and a random field $t \mapsto Y(t)$ as a mapping

$$T \cup T' \to L_{\mathbb{C}}^2(\Omega \times \Omega', \mathfrak{A} \otimes \mathfrak{A}', \mathbb{P} \otimes \mathbb{P}')$$

defined by

$$Y(t)(\omega, \omega') := \begin{cases} X(t)(\omega) & if \quad t \in T \\ X(t)(\omega') & if \quad t \in T'. \end{cases}$$

Applying the previously discussed case to the random field Y over $T \cup T'$ and identifying Y with its projection $t \mapsto X(t)$ from T into $L_{\mathbb{C}}^2(\Omega', \mathfrak{A}', \mathbb{P}')$ we obtain that

$$X(t) = \int_E a(t, x) Z(\cdot, dx)$$

whenever $t \in T$. Here,

$$H_{\mathbb{C}}^Z \cong L_{\mathbb{C}}^2(E, \mathfrak{E}, \mu)$$

and $H_{\mathbb{C}}^X \cong K$, hence $H_{\mathbb{C}}^X \underset{+}{\subset} H_{\mathbb{C}}^Z$. □

3. Stationarity of random fields

3.1. Graphs, buildings and their associated polynomial structures

3.1.1. Distance-transitive graphs and Cartier polynomials. Let Γ be an infinite graph which is locally finite in the sense that each vertex admits only finitely many neighbors. By $G := Aut(\Gamma)$ we denote the group of bijections $f : \Gamma \to \Gamma$ such that the images of neighbors remain neighbors under f. We assume that

(A) *G operates transitively on* G, i.e., for $s, t \in \Gamma$ there exists $g \in G$ such that $g(s) = t$.
Given $0\,(= t_0) \in \Gamma$ and the stabilizer $H := H_0$ of 0 one obtains a homeomorphism $gH \mapsto g(t_0)$ from G/H onto Γ. It follows that the double coset space $K := G/\!/H$ can be identified with the set Γ^H of orbits under the action of H on Γ. On the other hand it is shown that K is a discrete space admitting a convolution $*$ in $M^b(K)$ induced by the convolution in $M^b(G)$.
Now we make an additional assumption.

(B) Γ is *distance-transitive*, i.e. for the natural distance d in Γ defined for $s, t \in \Gamma$ as the minimal length $d(s,t)$ of paths connecting s and t, we require that for all $s, t, u, v \in \Gamma$ with $d(s,t) = d(u,v)$ there exists $g \in G$ satisfying $g(s) = u$ and $g(t) = v$.

It follows that H operates transitively on

$$\Gamma_n^H := \{t \in \Gamma : d(t,0) = n\} \ (n \in \mathbb{Z}_+).$$

In fact, every orbit in Γ^H is given by exactly one Γ_n^H. Next one shows that $K := G/\!/H$ is isomorphic to $\mathbb{Z}_+$, and the convolution $*$ in $M^b(K)$ carries over to $M^b(\mathbb{Z}_+)$ via

$$\varepsilon_m * \varepsilon_n = \sum_{k \geq 0} g(m,n,k)\varepsilon_k,$$

where the coefficients $g(m,n,k)$ are $\geq 0 \quad (m,n,k \in \mathbb{Z}_+)$. Defining the sequence $(Q_n)_{n \in \mathbb{Z}_+}$ of polynomials on $\mathbb{R}$ by

$$\begin{cases} Q_0 \equiv 1, \\ Q_1(x) = c_1 x + c_2 \quad \textit{for all} \quad x \in \mathbb{R}\ (c_1, c_2 \in \mathbb{R}, c_1 > 0), \quad \textit{and} \\ Q_1 Q_n = g(1,n,n+1)Q_{n+1} + g(1,n,n)Q_n + g(1,n,n-1)Q_{n-1}\ (n \geq 1), \end{cases}$$

Favard's theorem yields an orthogonal sequence admitting the nonnegative linearization

$$Q_m Q_n = \sum_{k \geq 0} g(m,n,k) Q_k \quad (m,n \in \mathbb{Z}_+). \tag{$*$}$$

Finally we apply MacPherson's theorem which says that under the assumptions (A) and (B) Γ is a *standard graph* $\Gamma(a,b)$ for all $a, b \geq 2$ with the defining property that at every vertex of Γ there are attached a copies of a complete graph with b vertices.

Applying this standardization $(Q_n)_{n \in \mathbb{Z}_+}$ appears to be the sequence $(Q_n^{a,b})$ of *Cartier polynomials* satisfying $(*)$ with coefficients involving only the numbers

a and b. These coefficients can be explicitly calculated, and they specify the convolution $*(Q_n^{a,b})$ in $M^b(\mathbb{Z}_+)$. The resulting convolution structure $(\mathbb{Z}_+, *(Q_n^{a,b})$ will be later identified as the *Cartier hypergroup* attached to the distance-transitive graph Γ.

For detailed arguments concerning the above discussion see, e.g., [5] and [13].

Special case. For the choices $a := q+1$ with $q \geq 1$ and $b := 2$ $(\mathbb{Z}_+, *(Q_n^{a,b}))$ becomes the *Arnaud-Dunau hypergroup* $(\mathbb{Z}_+, *(Q_n^q))$ attached to the homogeneous tree Γ_q of order q (see [4], D 1.1.7).

3.1.2. Triangle buildings and Cartwright polynomials. A triangle building can be viewed as a simplicial complex $\triangle$ consisting of vertices, edges and triangles such that any two triangles are connected by a gallery of triangles. It will be assumed that

(A) the graph formed by the vertices and edges of $\triangle$ is locally finite.

To every v in the set Γ of vertices of $\triangle$ there is associated its type $\tau(v) \in \{0,1,2\}$ by the property that any triangle has a vertex of each type.

In the sequel we shall restrict ourselves to *thick* triangular buildings $\triangle$ which enjoy the property that each edge of $\triangle$ belongs to the same number $q+1$ of triangles. q is said to be the order of $\triangle$. From now on we only sketch what has been detailed in the papers [11] and [7].

For $u, v \in \Gamma$ there exists a subcomplex A of $\triangle$ with $u, v \in \triangle$ such that A is isomorphic to a Euclidean plane tessellated by equilateral triangles. Moreover, there are unique elements $m, n \in \mathbb{Z}_+$ and rays $(u_0 = u, u_1, \ldots, u_n)$, $(u'_0 = u, u'_1, \ldots, u'_m)$ in A with $d(u_n, v) = m$, $d(u'_m, v) = n$ and $\tau(u_{i+1}) = \tau(u_i) + 1 \pmod 3$, $\tau(u'_{i+1}) = \tau(u'_i) - 1 \pmod 3$ $(i \geq 1)$ respectively. Since m, n are independent of A, for $u \in \Gamma$ and $m, n \in \mathbb{Z}_+$ the set $S_{n,m}(u)$ consisting of all $v \in \Gamma$ having the described properties is well defined and is such that $v \in S_{n,m}(u)$ if and only if $u \in S_{m,n}(v)$.

Let G denote the group $Aut_q(\Gamma)$ of automorphisms g of Γ which are *type-rotating* in the sense that there is a $c_g \in \{0,1,2\}$ with $\tau(g(u)) \equiv \tau(u) + c_g \ (mod\ 3)$ for all $u \in \Gamma$.

Then the stabilizer $H := H_0$ of any fixed $0 = v_0 \in \Gamma$ acts on $S_{n,m}(o)$ for every $n, m \in \mathbb{Z}_+$. Under the assumption that

(B) G acts transitively on Γ, and H acts transitively on each $S_{n,m}(o)$

we obtain the identifications

$$G//H \cong \{S_{n,m}(o) : n, m \in \mathbb{Z}_+\} \cong \mathbb{Z}_+^2.$$

Now, the method exposed in 3.1.1 yields a convolution $*$ in $M^b(\mathbb{Z}_+^2)$ described in terms of a sequence $(Q^q_{m,n})_{(m,n) \in \mathbb{Z}_+^2}$ of orthogonal polynomials on $\mathbb{C}^2$ admitting a nonnegative linearization analogous to $(*)$ of the form

$$Q^q_{m,n}\, Q^q_{k,l} = \sum_{(r,s) \in \mathbb{Z}_+^2} g((m,n),(k,l),(r,s)) Q^q_{r,s}.$$

The convolution structure $(\mathbb{Z}_+^2, *(Q_{m,n}^q))$ attached to the given triangular building $\triangle =: \triangle_q$ will later be made precise as the *Cartwright hypergroup of order* q.

For an explicit calculation of the Cartwright polynomials see [11].

Special case. Let q be a prime power and let F be a local field with residual field of order q and valuation $\mathfrak{v}$. We consider the set $Y := \{x \in F : \mathfrak{v}(x) \geq 0\}$ and fix $y \in Y$ with $\mathfrak{v}(y) = 1$. A lattice L in $V := F^3$ is any Y-submodule of V consisting of elements of the form $a_1v_1 + a_2v_2 + a_3v_3$ with $a_1, a_2, a_3 \in Y$, where $\{v_1, v_2, v_3\}$ is an F-basis of V. Lattices L, L' in V are said to be equivalent if $L' = wL$ for some $w \in F$. The equivalence classes [L] of lattices L in V form the set of vertices Γ_F of a building $\triangle_F$, whose triangles consist of distinct vertices $[L_1], [L_2], [L_3]$ satisfying the inclusions $L_1 \supset L_2 \supset L_3 \supset wL_1$. It can be shown that the group $G := PGL(3, F)$ acts on $\triangle_F$ by left multiplication as a group of type-rotating automorphisms, and the stabilizer H_0 of any vertex $0 = u_0$ of the graph Γ_F acts transitively on each of the sets $S_{n,m}(0)$ $(n, m \in \mathbb{Z}_+)$. For a proof of the latter fact see [6].

3.2. Stationary random fields over hypergroups

Hypergroups are locally compact spaces K admitting a continuous convolution $(\mu, \nu) \mapsto \mu * \nu$ in the Banach space $M^b(K)$ such that the convolution $\varepsilon_x * \varepsilon_y$ for $x, y \in K$ is a probability measure with compact (not necessarily singleton) support, and such that the axioms of. C.F. Dunkl, R.I. Jewett and R. Spector are satisfied. For the entire set of axioms and an extensive presentation of the theory see [4]. Here we just note that there is an involution $\mu \mapsto \mu^-$ in $M^b(K)$ and a neutral element ε_e for some distinguished $e \in K$. A hypergroup $(K, *)$ is called commutative provided $\varepsilon_x * \varepsilon_y = \varepsilon_y * \varepsilon_x$ for all $x, y \in K$.

Any analytic study on hypergroups relies on the notion of the *generalized translation* of an admissible function f on K try an element $x \in K$ defined by

$$f(x * y) := T^x f(y) := \int_K f(z)\varepsilon_x * \varepsilon_y(dz)$$

whenever $y \in K$.

Some analysis on a commutative hypergroup

1. On K there exists a unique translation invariant measure $\omega_K \in M_+(K)\backslash\{0\}$, called the *Haar measure* of K.
2. The set of *characters* of K consisting of all continuous mappings $\chi : K \to \mathbb{D} := \{z \in \mathbb{C} :| z |\leq 1\}$ such that

$$T^x\chi(y^-) = \chi(x)\overline{\chi(y)}$$

for all $x, y \in K$, together with the compact open topology becomes a locally compact space $K^\wedge$, the *dual* of K.

3. There is an injective *Fourier-Stiehltjes transform* $\mu \mapsto \widehat{\mu}$ on $M^b(K)$, where $\widehat{\mu}$ is given by

$$\widehat{\mu}(\chi) := \int \overline{\chi(x)}\mu(dx)$$

whenever $\chi \in K^\wedge$.

4. The Fourier transform $f \mapsto \widehat{f}$ establishes an isometric isomorphism

$$L^2(K, \omega_K) \hookrightarrow L^2(K^\wedge, \pi_K),$$

where the measure $\pi_K \in M_+(K^\wedge)$ serves as the *Plancherel measure* associated with the Haar measure ω_K of K.

One observes that in general $K^\wedge$ is not a hypergroup (with respect to pointwise composition), and that the support of π_K can be strictly smaller than $K^\wedge$.

Clearly, any locally compact group is a hypergroup, and in the case of Abelian groups properties 2. to 4. do not present the obstacles mentioned.

Examples of commutative hypergroups

(1) If (G, H) is a *Gelfand pair* in the sense that the algebra of H-biinvariant functions in $L^1(G, \mathfrak{B}(G), \omega_G)$ (ω_G denoting the Haar measure of G) is commutative, then $K := G//H$ is a commutative hypergroup $(K, *)$ with convolution $*$ given by

$$\varepsilon_{HxH} * \varepsilon_{HyH} = \int_H \varepsilon_{HxhyH}\, \omega_H(dh),$$

involution $HxH \mapsto Hx^{-1}H$, neutral element $H = HeH$ (e denoting the neutral element of G) and Haar measure

$$\omega_K = \int_G \varepsilon_{HxH}\, \omega_G(dx).$$

(1.1) The choice (G, H) with $H = \{e\}$ yields the case $K = G$ of an Abelian locally compact group.

(1.2) In 3.1.1 we introduced the Cartier hypergroup $(\mathbb{Z}_+, *(Q_n^{a,b}))$ as $(K, *)$ with $K := G//H$ and $G/H \cong \Gamma$. Its Haar measure is given by

$$\omega_k(\{k\}) = \frac{1}{g(k,k,0)} = a(a-1)^{k-1}(b-1)^k$$

for all $k \in \mathbb{Z}_+$. Moreover,

$$K^\wedge \cong \{x \in \mathbb{R} : (Q_n^{a,b}(x)) \text{ is bounded }\}$$

can be identified with the interval $[-s_1, s_1]$, where s_1 satisfying the inequalities $-s_1 \leq -1 < 1 \leq s_1$ can be computed in terms of a and b, and the Plancherel measure π_K admits a density with respect to Lebesgue measure λ restricted to $[-1, 1]$ (depending on a, b).

(1.3) For the Cartwright hypergroup $(\mathbb{Z}_+^2, *(Q_{m,n}^q))$ of 2.1.2 we have the identification with $(K, *)$, where $K := G//H$ and G/H is isomorphic to the graph Γ attached to the given triangle building. One notes that the involution of K is

$(m,n) \mapsto (n,m)$ and the neutral element is just $(0,0)$. While Haar measure ω_K of K is easily computed as

$$\omega_K(\{(m,n)\}) = \begin{cases} 1 & \text{if} \quad m = n = 0 \\ (q^2+q+1)q^{2(m-1)} & \text{if} \quad m = 0 \text{ or } n = 0 \\ (q^2+q+1)(q^2+q)q^{2(m+n-2)} & \text{if} \quad m,n \geq 1, \end{cases}$$

the description of $K^\wedge$ and of π_K requires more effort. In fact,

$$K^\wedge = \{(z,\overline{z}) \mapsto Q^q_{m,n}(z,\overline{z}) : z \in Z_q\}$$

can be identified with the compact region Z_q in $\mathbb{C}$ bounded by the closed Jordan curve $t \mapsto e^{zit} + (q + \frac{1}{q})e^{it}$ $(t \in [0, 2\pi])$, and the Plancherel measure π_K has a density (depending on q) with respect to λ^2 restricted to Z (see [11]).

It should be noted that the hypergroups in Examples (1.2) and (1.3) are special types of *polynomial hypergroups* $(\mathbb{Z}^d_+, *)$ in *d variables* the defining sequences of polynomials being interpreted as the spherical functions of the corresponding Gelfand pairs (see [14] and [12]).

With this portion of knowledge of hypergroup theory we are going to study random fields X over $T := K$ with mean function m on K and covariance kernel C on $K \times K$.

A centered random field X over K is called *stationary* if $C \in C^b(K \times K)$ and

$$C(a,b) = T^a C(b^-, a) = \int_K C(x,e)\varepsilon_a * \varepsilon_{b^-}(dx)$$

for all $a, b \in K$. The function $C_1 := C(\cdot, e)$ is said to be the *covariance function* of X.

From the general construction layed out in Section 2.1 we know that for any $C \in C^b(K)$ the positive definiteness of C is equivalent to the existence of a stationary random field over K with covariance function C. In view of the theorem in Section 2.3 we have the following result quoted from [4], Section 8.2.

Theorem. Let X be any random field over a commutative hypergroup K with covariance kernel C. Then the following statements are equivalent:

(i) X is stationary.

(ii) (Spectral representation). There exists a *spectral measure* $\mu \in M^b_+(K^\wedge)$ such that

$$C(a,b) = \int_{K^\wedge} \chi(a)\overline{\chi(b)}\mu(d\chi)$$

for all $a, b \in K$.

(iii) (Karhunen representation). There exists a centered spatial random field $A \mapsto Z(A)$ over $T := \mathfrak{B}(K^\wedge)$ with orthogonal increments and base $\mu \in M^b_+(K^\wedge)$ such that

$$X(a) = \int_{K^\wedge} \chi(a) Z(d\chi) \quad \text{for all } a \in K.$$

From (iii) follows that

$$\| Z(A) \|_2^2 = \mu(A) \quad \text{whenever } A \in T.$$

The theorem applies to double coset spaces arising from Gelfand pairs (G, H), in particular to the following

Special cases

(1) $G := Aut(\Gamma)$ for a distance-transitive graph $\Gamma := \Gamma(a, b)$ and $H := H_0$ for some $0 := t_0 \in \Gamma$ (see 3.1.1).
(2) $G := Aut_q(\Gamma)$ for the graph Γ of vertices of a triangle building $\triangle_q$ and $H := H_0$ for some $0 := v_0 \in \Gamma$ (see 3.1.2).

3.3. Arnaud-Letac stationarity

Let $\Gamma := \Gamma_q$ denote a homogeneous tree of order $q \geq 1$ (see 3.1.1, Special case). A centered random field X over $T := \Gamma$ with covariance kernel C is called *stationary* (in the sense of Arnaud-Letac) if there exists a mapping $\varphi : \to \mathbb{R}$ such that

$$C(s,t) = \varphi(d(s,t))$$

where $s, t \in \Gamma$.

Clearly, the distribution of X (as a measure on the path space $\mathbb{C}^\Gamma$) is invariant under the action of $Aut(\Gamma)$.

Given a stationary random field X over Γ we want to establish an isometry

$$H_{\mathbb{C}}^X(\Gamma) := \langle \{X(t) : t \in \Gamma\} \rangle^- \to L_{\mathbb{C}}^2(E, \mathfrak{E}, \mu)$$

for some measure space $(E, \mathfrak{E}, \mu)$, where μ is related to the real spectral measure μ_X of X. In order to motivate the construction due to J.-P. Arnaud ([2]) we refer to two very **special situations** (see [3]).

(1) For $q = 1$ Karhunen's approach yields the choices $E := \mathbb{T} := \mathbb{R}/\mathbb{Z}, \mathfrak{E} := \mathfrak{B}(\mathbb{T})$ and μ equal to the spectral measure μ_X of the random field X over $\mathbb{Z}$. The desired isometry is given by

$$X(n) \mapsto \chi_n,$$

where $\vartheta \mapsto \chi_n(\vartheta)$ is the character of $\mathbb{T}$ defined for $n \in \mathbb{Z}$. The corresponding spectral representation reads as

$$C(X(m), X(n)) = \int_{\mathbb{T}} e^{im\vartheta} e^{-in\vartheta} \mu(d\vartheta)$$

valid for all $m, n \in \mathbb{Z}$.

We observe in this special situation that the distribution of X is invariant under the *commutative* group of translations of $\mathbb{T}$. The situation of a noncommutative invariance group as it appears, e.g., for Riemannian symmetric pairs (G, H) is more difficult to handle as can be seen in the case of the homogeneous space $G/H = \mathbb{S}^2 := \{t \in \mathbb{R}^3 : || t || = 1\}$.

(2) Let X be a centered random field over $\mathbb{S}^2$ with state space $\mathbb{R}$ having covariance kernel C. We assume that there exists a continuous function $\vartheta : I := [-1, 1] \to \mathbb{R}$ satisfying

$$C(s,t) = \vartheta(\langle s, t \rangle)$$

for all $s, t \in \mathbb{S}^2$. Such functions φ are necessarily of the form

$$\varphi(x) = \sum_{n \geq 0} a_n(\varphi) Q_n(x) \ (x \in I),$$

where $(Q_n)_{n \in \mathbb{Z}_+}$ denotes the sequence of Legendre polynomials on I and $(a_n(\varphi))_{n \in \mathbb{Z}_+}$ is a sequence in $\mathbb{R}$ such that $\sum_{n \geq 0} a_n(\varphi) < \infty$. Now we introduce for every $n \in \mathbb{Z}_+$ the space E_n of eigenvectors corresponding to the eigenvalue $-n(n+1)$ of the Laplacian on $\mathbb{S}^2$. Clearly, $dim E_n = 2n+1$. By $\{\triangle_{k,n} : k \in \{-n, \ldots, n\}\}$ we denote an orthonormal basis of E_n and consider E_n as a subspace of $L^2_{\mathbb{R}}(\mathbb{S}^2, \mathfrak{B}(\mathbb{S}^2), \sigma)$, where σ is the unique rotation invariant measure on $\mathbb{S}^2$. Then

$$Q_n(\langle s, t \rangle) = \frac{1}{2n+1} \sum_{k=-n}^{n} \triangle_{k,n}(s) \overline{\triangle_{k,n}(t)}$$

whenever $s, t \in \mathbb{S}^2$.

Now, let $(Z_{k,n})_{k \in \{-n,\ldots,n\}}$ be a sequence of independent $N(0,1)$-distributed random variables on $(\Omega, \mathfrak{A}, \mathbb{P})$. Then

$$Y(t) := \sum_{n \geq 0} \sqrt{\frac{a_n(\varphi)}{2n+1}} \sum_{k=-n}^{n} Z_{k,n} \triangle_{k,n}(t) \ (t \in \mathbb{S}^2)$$

defines a random field Y over $\mathbb{S}^2$ with covariance kernel C given by $\varphi(\langle \cdot, \cdot \rangle)$. Next, let

$$E := \{(k,n) \in \mathbb{Z} \times \mathbb{Z}_+ : \mid k \mid \leq n\}$$

and

$$\mu_\varphi(\{(k,n)\}) := \frac{a_n(\varphi)}{2n+1}$$

for all $(k,n) \in E$. Finally, for fixed $t \in \mathbb{S}^2$ let $\triangle_t$ denote the mapping $(k,n) \mapsto \triangle_{k,n}(t)$ from E into $\mathbb{C}$. Then

$$\int_E \triangle_s \overline{\triangle_t} \, d\mu_\varphi = \varphi(\langle s, t \rangle)$$

for all $s, t \in \mathbb{S}^2$. Linear extension of $Y(t) \mapsto \triangle_t$ yields the desired isometry

$$H^Y_{\mathbb{C}}(\mathbb{S}^2) \to L^2_{\mathbb{C}}(E, \mathfrak{P}(E), \mu_\varphi)$$

($\mathfrak{P}(E)$ denoting the power set of E.)

Returning to the initially given homogenous tree $\Gamma := \Gamma_q$ of order $q > 1$ we at first state a result analogous to the equivalence (i) $\Leftrightarrow$ (ii) of the theorem in 3.2.

Theorem. (Spectral representation, [1]) For any centered random field X over Γ with covariance kernel C the following statements are equivalent:

(i) X is stationary.
(ii) There exists a unique spectral measure $\mu_X \in M_+^b(I)$ such that

$$C(s,t) = \int_I Q_{d(s,t)}(x)\mu_X(dx)$$

whenever $s, t \in \Gamma$. Here $(Q_n)_{n \in \mathbb{Z}_+}$ denotes the sequence of Arnaud-Dunau polynomials $Q_n := Q_n^q$ $(n \in \mathbb{Z}_+)$.

The announced Karhunen isometry is achieved in the subsequent

Theorem. (Karhunen representation, [2]). Let X be a stationary random field over Γ with covariance kernel C and spectral measure Γ_X. Then for all $s, t \in \Gamma$ one has

$$C(s,t) = \int_E \triangle_s(\xi)\overline{\triangle_t(\xi)}\mu(d\xi).$$

Here $(E, \mathfrak{E}, \mu)$ denotes a measure consisting of the Cantor sphere

$$E := (I^0 \times B) \cup \{-1, 1\}$$

of order q, where B is the compact space of ends of Γ, $\mathfrak{E}$ its Borel σ-algebra $\mathfrak{B}(E)$ and the measure

$$\mu := \mu_X \otimes \nu_0 \in M_+^b(E),$$

where ν_0 is the unique probability measure on B invariant under the compact group of isometries stabilizing an a priori fixed vertex $0 := t_0 \in \Gamma$, with

$$\mu(\{-1\}) = \mu_X(\{-1\}), \qquad \mu(\{1\}) = \mu_X(\{1\}),$$

and $(\triangle_t)_{t \in \Gamma}$ denotes a set of functions $E \to \mathbb{C}$ replacing the characters in the hypergroup case.

Moreover, the mapping $X(t) \mapsto \triangle_t$ provides an isometry

$$H_{\mathbb{C}}^X(\Gamma) \to L_{\mathbb{C}}^2(E, \mathfrak{E}, \mu).$$

Special case. If the spectral measure μ_X of the random field X over Γ is concentrated on the interval $I_q := [-\rho_q, \rho_q]$ with $\rho_q := \frac{2\sqrt{q}}{q+1}$, then

$$Q_{d(s,t)}(x) = \int_B \triangle_t(x,b)\overline{\triangle_s(x,b)}\nu_0(db)$$

whenever $x \in I_q$, the functions $\triangle_t : I_q \times B \to \mathbb{C}$ being of the form

$$\triangle_t(x,b) := (\sqrt{q} \exp i\,\vartheta)^{\delta_b(0,t)}$$

with $\vartheta := \arccos(\frac{x}{\delta_q})$ and

$$\delta_b(s,t) := \lim_{u \to b}(d(u,s) - d(u,t))\ (s, t \in \Gamma).$$

Since

$$C(s,t) = \int_{I_q} Q_{d(s,t)}(x)\mu_X(dx) = \int_{I_q \times B} \triangle_t(x,b)\overline{\triangle_s(x,b)}\mu_X(dx)\nu_0(db)$$

for all $s, t \in \Gamma$, the mapping $X(t) \mapsto \triangle_t$ is an isometry

$$H_{\mathbb{C}}^X(\Gamma) \to L^2(I_q \times B, \mathfrak{B}(I_q \times B), \mu_X \otimes \nu_0).$$

References

[1] J.-P. Arnaud: Fonctions sphériques et fonctions définies positives sur l'arbre homogène. C. R. Acad. Sci. Paris 290, Série A (1980) 99–101

[2] J.-P. Arnaud: Stationary processes indexed by a homogeneous tree. The Annals of Probability Vol. 22, no. 1 (1994), 195–218

[3] J.-P. Arnaud, G. Letac: La formule de représentation spectrale d'un processus gaussien stationnaire sur un arbre homogène. Publication du Laboratoire de Statistique et Probabilités de l'Université Paul Sabatier, no. 1 (1985), 1–11

[4] W.R. Bloom, H. Heyer: Harmonic Analysis of Probability Measures on Hypergroups. Walter de Gruyter 1995

[5] P. Cartier: Harmonic analysis on trees. Proc. Symp. Pure Math. 26 (1976), 419–424

[6] D.I. Cartwright, W. Mlotkowski: Harmonic analysis for groups acting on triangle buildings. J. Austral. Math. Soc. 56 (1994), 345–383

[7] D.I. Cartwright, W. Woess: Isotropic random walks in a building of type $\tilde{A}_d$. Math. Z. 247 (2004), 101–135

[8] D. Dacunha-Castelle, M. Duflo: Probability and Statistics. Volume II. Springer-Verlag 1986

[9] R. Gangolli: Positive definite kernels on homogeneous spaces and certain stochastic processes related to Lévy's Brownian motion of serveral parameters. Ann. Inst. H. Poincaré Sect. B (N.S.) 3 (1967), 121–226

[10] R. Lasser, M. Leitner: Stochastic processes indexed by hypergroups I. J. Theoret. Probab. 2, no. 3 (1989), 301–311

[11] M. Lindlbauer, M. Voit: Limit theorems for isotropic random walks on triangle buildings. J. Austral. Math. Soc. 73 (2002), 301–333

[12] A.M. Mantero, A. Zappa: Spherical functions and spectrum of the Laplace operators on buildings of rank 2. Boll. Unione Mat. Ital. VII Sér., B 8 (1994), 419–475

[13] M. Voit: Central limit theorems for random walks on ${}_0$ that are associated with orthogonal polynomials. J. Multivariate Analysis 34 (1990), 290–322

[14] Hm. Zeuner: Polynomial hypergroups in several variables. Arch. Math. 58 (1992), 425–434

Herbert Heyer
Mathematisches Institut
Eberhard-Karls-Universität Tübingen
Auf der Morgenstelle 10
D-72076 Tübingen, Germany
e-mail: herbert.heyer@uni-tuebingen.de

Operator Theory:
Advances and Applications, Vol. 167, 173–182

Matrix Representations and Numerical Computations of Wavelet Multipliers

M.W. Wong and Hongmei Zhu

Abstract. Weyl-Heisenberg frames are used to obtain matrix representations of wavelet multipliers. Simple test cases are used to illustrate the numerical computations of wavelet multipliers by means of the corresponding truncated matrices.

Mathematics Subject Classification (2000). Primary: 47G10, 47G30, 65T50, 65T60.

Keywords. Wavelet multipliers, Weyl-Heisenberg frames, admissible wavelets.

1. Wavelet multipliers

Let π be the unitary representation of the additive group $\mathbb{R}^n$ on $L^2(\mathbb{R}^n)$ defined by

$$(\pi(\xi)u)(x) = e^{ix\cdot\xi}u(x), \quad x, \xi \in \mathbb{R}^n,$$

for all functions u in $L^2(\mathbb{R}^n)$. The following two results can be found in the paper [4] by He and Wong.

Theorem 1.1. *Let $\varphi \in L^2(\mathbb{R}^n) \cap L^\infty(\mathbb{R}^n)$ be such that $\|\varphi\|_{L^2(\mathbb{R}^n)} = 1$. Then for all functions u and v in the Schwartz space $\mathcal{S}$,*

$$(2\pi)^{-n} \int_{\mathbb{R}^n} (u, \pi(\xi)\varphi)_{L^2(\mathbb{R}^n)} (\pi(\xi)\varphi, v)_{L^2(\mathbb{R}^n)} d\xi = (\varphi u, \varphi v)_{L^2(\mathbb{R}^n)}.$$

Theorem 1.2. *Let $\sigma \in L^\infty(\mathbb{R}^n)$ and let φ be any function in $L^2(\mathbb{R}^n) \cap L^\infty(\mathbb{R}^n)$ such that $\|\varphi\|_{L^2(\mathbb{R}^n)} = 1$. If, for all functions u in $\mathcal{S}$, we define $P_{\sigma,\varphi}u$ by*

$$(P_{\sigma,\varphi}u, v)_{L^2(\mathbb{R}^n)} = (2\pi)^{-n} \int_{\mathbb{R}^n} \sigma(\xi)(u, \pi(\xi)\varphi)_{L^2(\mathbb{R}^n)} (\pi(\xi)\varphi, v)_{L^2(\mathbb{R}^n)} d\xi, \quad v \in \mathcal{S}, \tag{1.1}$$

This research has been supported by the Natural Sciences and Engineering Research Council of Canada.

then

$$(P_{\sigma,\varphi}u, v)_{L^2(\mathbb{R}^n)} = ((\varphi T_\sigma \overline{\varphi})u, v)_{L^2(\mathbb{R}^n)}, \quad u, v \in \mathcal{S}, \tag{1.2}$$

where T_σ *is the pseudo-differential operator with constant coefficients or Fourier multiplier given by*

$$(T_\sigma u)(x) = (2\pi)^{-n/2} \int_{\mathbb{R}^n} e^{ix\cdot\xi} \sigma(\xi)\hat{u}(\xi)\, d\xi, \quad u \in \mathcal{S}. \tag{1.3}$$

The Fourier transform $\hat{u}$ of u in (1.3), also denoted by $\mathcal{F}u$, is chosen to be that defined by

$$\hat{u}(\xi) = (2\pi)^{-n/2} \int_{\mathbb{R}^n} e^{-ix\cdot\xi} u(x)\, dx, \quad \xi \in \mathbb{R}^n.$$

The inverse Fourier transform $\check{v}$ of a function v in $\mathcal{S}$, also denoted by $\mathcal{F}^{-1}v$, is then given by

$$\check{v}(x) = (2\pi)^{-n/2} \int_{\mathbb{R}^n} e^{ix\cdot\xi} v(\xi)\, d\xi, \quad x \in \mathbb{R}^n.$$

Let $\sigma \in L^\infty(\mathbb{R}^n)$. Then the linear operator $P_{\sigma,\varphi} : L^2(\mathbb{R}^n) \to L^2(\mathbb{R}^n)$, defined by (1.2) for all functions u and v in $\mathcal{S}$, is in fact a bounded linear operator, and can be considered as a variant of a localization operator corresponding to the symbol σ and the admissible wavelet φ studied in the book [12] by Wong. Had the "admissible wavelet" φ been replaced by the function φ_0 on $\mathbb{R}^n$ given by

$$\varphi_0(x) = 1, \quad x \in \mathbb{R}^n,$$

the equation (1.1) would have been a resolution of the identity formula and the operator P_{σ,φ_0} would have been a usual pseudo-differential operator with constant coefficients or a Fourier multiplier. This is the rationale for calling the linear operator $P_{\sigma,\varphi} : L^2(\mathbb{R}^n) \to L^2(\mathbb{R}^n)$ a wavelet multiplier.

Wavelet multipliers arise naturally as filters in signal analysis and have been studied extensively and systematically in Chapters 19–21 of the book [12] by Wong, the paper [13] by Wong and Zhang, and the Ph.D. dissertation [14] by Zhang.

We need in this paper another representation of the wavelet multiplier $\varphi T_\sigma \overline{\varphi}$ as a Hilbert-Schmidt operator on $L^2(\mathbb{R}^n)$.

Theorem 1.3. *Let* $\sigma \in L^1(\mathbb{R}^n) \cap L^2(\mathbb{R}^n) \cap L^\infty(\mathbb{R}^n)$ *and let* $\varphi \in L^2(\mathbb{R}^n) \cap L^\infty(\mathbb{R}^n)$. *If we let* K *be the function on* $\mathbb{R}^n \times \mathbb{R}^n$ *defined by*

$$K(x, y) = \varphi(x)k(x - y)\overline{\varphi}(y), \quad x, y \in \mathbb{R}^n,$$

where

$$k = (2\pi)^{-n/2}\check{\sigma},$$

then $K \in L^2(\mathbb{R}^n \times \mathbb{R}^n)$ *and*

$$((\varphi T_\sigma \overline{\varphi})f)(x) = \int_{\mathbb{R}^n} K(x, y)\, f(y)\, dy, \quad x \in \mathbb{R}^n.$$

In particular, $\varphi T_\sigma \overline{\varphi}$ *is a Hilbert-Schmidt operator.*

Proof. Since $k \in L^\infty(\mathbb{R}^n)$,

$$\begin{aligned}\int_{\mathbb{R}^n}\int_{\mathbb{R}^n}|K(x,y)|^2 dx\,dy &= \int_{\mathbb{R}^n}\int_{\mathbb{R}^n}|\varphi(x)|^2|k(x-y)|^2|\varphi(y)|^2 dx\,dy \\ &\leq \|k\|^2_{L^\infty(\mathbb{R}^n)}\|\varphi\|^4_{L^2(\mathbb{R}^n)} < \infty.\end{aligned}$$

Now, for all functions f in $L^2(\mathbb{R}^n)$,

$$(k * f)^\wedge = (2\pi)^{n/2}\hat{k}\hat{f} = \sigma\hat{f}$$

and hence

$$k * f = (\sigma\hat{f})^\vee = T_\sigma f.$$

Therefore for all functions in $L^2(\mathbb{R}^n)$, we get

$$\begin{aligned}((\varphi T_\sigma \overline{\varphi})f)(x) &= \varphi(x)(k * (\overline{\varphi}f))(x) \\ &= \int_{\mathbb{R}^n}\varphi(x)k(x-y)\overline{\varphi(y)}\,f(y)\,dy \\ &= \int_{\mathbb{R}^n}K(x,y)\,f(y)\,dy\end{aligned}$$

for all x in $\mathbb{R}^n$. □

The motivation for this paper comes from a desire to develop "efficient" numerical methods for the computations of pseudo-differential operators studied in the books [10, 11] by Wong, among others. As wavelet multipliers are very simple, but not trivial, pseudo-differential operators, they serve very well as test cases for any proposed numerical methods. The aim of this paper is to obtain matrix representations of wavelet multipliers using Theorem 1.3 and Weyl-Heisenberg frames. We first show in Section 2 that wavelet multipliers are natural and important pseudo-differential operators in their own right by recalling succinctly the fact in the book [12] to the effect that the well-known Landau-Pollak-Slepian operator is a wavelet multiplier. In Section 3, we give a brief recapitulation on the general theory of frames in Hilbert spaces and then we reconstruct the Weyl-Heisenberg frame, which is used in Section 4 to obtain matrix representations of wavelet multipliers. Some simple test cases are then used in Section 5 to illustrate the numerical computations of wavelet multipliers by implementing MATLAB 7.0 on the corresponding truncated matrices.

2. The Landau-Pollak-Slepian operator

Let Ω and T be positive numbers. Then we define the linear operators $P_\Omega : L^2(\mathbb{R}^n) \to L^2(\mathbb{R}^n)$ and $Q_T : L^2(\mathbb{R}^n) \to L^2(\mathbb{R}^n)$ by

$$(P_\Omega f)^\wedge(\xi) = \begin{cases} \hat{f}(\xi), & |\xi| \leq \Omega, \\ 0, & |\xi| > \Omega, \end{cases}$$

and

$$(Q_T f)(x) = \begin{cases} f(x), & |x| \leq T, \\ 0, & |x| > T, \end{cases}$$

for all functions f in $L^2(\mathbb{R}^n)$. Then it is easy to see that $P_\Omega : L^2(\mathbb{R}^n) \to L^2(\mathbb{R}^n)$ and $Q_T : L^2(\mathbb{R}^n) \to L^2(\mathbb{R}^n)$ are self-adjoint projections. In signal analysis, a signal is a function f in $L^2(\mathbb{R}^n)$. Thus, for all functions f in $L^2(\mathbb{R}^n)$, the function $Q_T P_\Omega f$ can be considered to be a filtered signal. Therefore it is of interest to compare the energy $\|Q_T P_\Omega f\|^2_{L^2(\mathbb{R}^n)}$ of the filtered signal $Q_T P_\Omega f$ with the energy $\|f\|^2_{L^2(\mathbb{R}^n)}$ of the original signal f. In fact, we can prove that

$$\sup \left\{ \frac{\|Q_T P_\Omega f\|^2_{L^2(\mathbb{R}^n)}}{\|f\|^2_{L^2(\mathbb{R}^n)}} : f \in L^2(\mathbb{R}^n),\ f \neq 0 \right\} = \|P_\Omega Q_T P_\Omega\|_*,$$

where $\| \ \|_*$ is the norm in the C^*-algebra of all bounded linear operators on $L^2(\mathbb{R}^n)$.

The bounded linear operator $P_\Omega Q_T P_\Omega : L^2(\mathbb{R}^n) \to L^2(\mathbb{R}^n)$ alluded to in the context of filtered signals is called the Landau-Pollak-Slepian operator. See the fundamental papers [5, 6] by Landau and Pollak, [7, 8] by Slepian and [9] by Slepian and Pollak for more detailed information.

The following theorem tells us that the Landau-Pollak-Slepian operator is in fact a wavelet multiplier.

Theorem 2.1. *Let φ be the function on $\mathbb{R}^n$ defined by*

$$\varphi(x) = \begin{cases} \frac{1}{\sqrt{\mu(B_\Omega)}}, & |x| \leq \Omega, \\ 0, & |x| > \Omega, \end{cases}$$

where $\mu(B_\Omega)$ is the volume of B_Ω, and let σ be the characteristic function on B_T, i.e,

$$\sigma(\xi) = \begin{cases} 1, & |\xi| \leq T, \\ 0, & |\xi| > T. \end{cases}$$

Then the Landau-Pollak-Slepian operator $P_\Omega Q_T P_\Omega : L^2(\mathbb{R}^n) \to L^2(\mathbb{R}^n)$ is unitarily equivalent to a scalar multiple of the wavelet multiplier $\varphi T_\sigma \varphi : L^2(\mathbb{R}^n) \to L^2(\mathbb{R}^n)$. In fact,

$$P_\Omega Q_T P_\Omega = \mu(B_\Omega)\mathcal{F}^{-1}(\varphi T_\sigma \varphi)\mathcal{F}.$$

Details on the Landau-Pollak-Slepian operator can be found in Chapter 20 of the book [12] by Wong.

3. Frames in Hilbert spaces

Frames, usually attributed to Duffin and Schaeffer [3] in the context of nonharmonic series, are alternatives to orthonormal bases for Hilbert spaces.

Let $\{x_j\}_{j=1}^{\infty}$ be a sequence in a separable, complex and infinite-dimensional Hilbert space X. We denote the inner product and norm in X by $(\ ,\ )$ and $\|\ \|$ respectively. Suppose that there exist positive constants A and B such that

$$A\|x\|^2 \leq \sum_{j=1}^{\infty} |(x, x_j)|^2 \leq B\|x\|^2, \quad x \in X.$$

Then we call $\{x_j\}_{j=1}^{\infty}$ a frame for X with frame bounds A, B with the understanding that A is the lower frame bound and B is the upper frame bound. The frame is said to be tight if $A = B$. The frame is called an exact frame if it is no longer a frame whenever an element is removed from it.

Of particular relevance for us in this paper are tight frames with frame bounds 1,1. This is due to the following theorem.

Theorem 3.1. *Let $\{x_j\}_{j=1}^{\infty}$ be a tight frame with frame bounds* 1,1. *Then*

$$x = \sum_{j=1}^{\infty} (x, x_j)x_j, \quad x \in X.$$

It is clear from Theorem 3.1 that a tight frame with frame bounds 1,1 provides an expansion, which is formally the same as that given by an orthonormal basis.

Frames and their applications abound in the literature. See, for instance, the book [1] by Christensen in this connection.

We need the following result in the paper [2] by Daubechies, Grossmann and Meyer. Since the proof is not difficult, we include a proof for the sake of being self-contained.

Theorem 3.2. *Let $g \in L^2(\mathbb{R})$ and let a and b be positive numbers such that we can find positive numbers A and B for which*

$$A \leq \sum_{n\in\mathbb{Z}} |g(x + na)|^2 \leq B \tag{3.1}$$

for almost all x in $\mathbb{R}$, and

$$\operatorname{supp}(g) \subseteq I,$$

where I is some interval with length $\frac{2\pi}{b}$. Then $\{M_{mb}T_{na}g : m, n \in \mathbb{Z}\}$ is a frame for $L^2(\mathbb{R})$ with frame bounds $\frac{2\pi A}{b}$ and $\frac{2\pi B}{b}$.

Proof. Let $n \in \mathbb{Z}$. Then

$$\operatorname{supp}(f(T_{na}\overline{g})) \subseteq I_n,$$

where $I_n = \{x - na : x \in I\}$. By (3.1), $g \in L^\infty(\mathbb{R})$. So, $f(T_{na}\overline{g}) \in L^2(I_n)$. Now, $\left\{\sqrt{\frac{b}{2\pi}} E_{mb} : m \in \mathbb{Z}\right\}$ is an orthonormal basis for $L^2(I_n)$, where

$$E_{mb}(x) = e^{imbx}, \quad x \in \mathbb{R}.$$

So,

$$\sum_{m\in\mathbb{Z}}\left|\left(f(T_{na}\overline{g}),\sqrt{\frac{b}{2\pi}}E_{mb}\right)_{L^2(\mathbb{R})}\right|^2=\|f(T_{na}\overline{g})\|^2_{L^2(\mathbb{R})}=\int_{-\infty}^{\infty}|f(x)|^2|g(x+na)|^2dx.$$

Therefore

$$\sum_{m\in\mathbb{Z}}\left|(f(T_{na}\overline{g}),E_m)_{L^2(\mathbb{R})}\right|^2=\frac{2\pi}{b}\int_{-\infty}^{\infty}|f(x)|^2|g(x+na)|^2dx.$$

Thus,

$$\begin{aligned}\sum_{m,n\in\mathbb{Z}}\left|(f,M_{mb}T_{na}g)_{L^2(\mathbb{R})}\right|^2 &= \sum_{m,n\in\mathbb{Z}}\left|(f(T_{na}\overline{g}),E_{mb})_{L^2(\mathbb{R})}\right|^2\\ &= \sum_{n\in\mathbb{Z}}\sum_{m\in\mathbb{Z}}\left|(f(T_{na}\overline{g}),E_{mb})_{L^2(\mathbb{R})}\right|^2\\ &= \frac{2\pi}{b}\sum_{n\in\mathbb{Z}}\int_{-\infty}^{\infty}|f(x)|^2\,|g(x+na)|^2dx\\ &= \frac{2\pi}{b}\int_{-\infty}^{\infty}|f(x)|^2\sum_{n\in\mathbb{Z}}|g(x+na)|^2dx.\end{aligned}$$

So, by (3.1),

$$\frac{2\pi A}{b}\|f\|^2_{L^2(\mathbb{R})}\le\sum_{m,n\in\mathbb{Z}}\left|(f,M_{mb}T_{na}g)_{L^2(\mathbb{R})}\right|^2\le\frac{2\pi B}{b}\|f\|^2_{L^2(\mathbb{R})}$$

and the proof is complete. □

Corollary 3.3. *Let $g\in L^2(\mathbb{R})$ be such that*

$$\sum_{n\in\mathbb{Z}}|g(x+2\pi n)|^2=1$$

for almost all x in $\mathbb{R}$ and

$$\operatorname{supp}(g)\subseteq[0,1].$$

Then $\{M_{2\pi m}T_{2\pi n}g : m,n\in\mathbb{Z}\}$ is a tight frame for $L^2(\mathbb{R})$ with frame bounds 1,1.

Remark 3.4. The frames in Theorem 3.2 and Corollary 3.3 are generated from one single function g in $L^2(\mathbb{R})$ by translations and modulations. As such, they are intimately related to the Weyl-Heisenberg group in signal analysis and hence they are known as Weyl-Heisenberg frames. The functions g generating the frames are called the admissible windows or wavelets. An admissible wavelet g for the tight frame in Corollary 3.3 can be given by $g=\chi_{[0,1]}$.

4. Matrix representations of wavelet multipliers

Let $g \in L^2(\mathbb{R})$ be as in Corollary 3.3. Let $g_{mn} = M_{2\pi m}T_{2\pi n}g$, where $m, n \in \mathbb{Z}$. Then $\{g_{mn} : m, n \in \mathbb{Z}\}$ is a tight frame for $L^2(\mathbb{R})$ with frame bounds 1,1. For all integers m, n, μ and ν, we let $G_{mn\mu\nu}$ be the tensor product of g_{mn} and $g_{\mu\nu}$, *i.e.*,

$$G_{mn\mu\nu}(x, y) = g_{mn}(x)g_{\mu\nu}(y), \quad m, n, \mu, \nu \in \mathbb{Z}.$$

Then we have the following proposition.

Proposition 4.1. *$\{G_{mn\mu\nu} : m, n, \mu, \nu \in \mathbb{Z}\}$ is a tight frame for $L^2(\mathbb{R}^2)$ with frame bounds* 1,1.

Proof. Let $F \in L^2(\mathbb{R}^2)$. Then

$$\begin{aligned}
&\sum_{m,n,\mu,\nu\in\mathbb{Z}} \left|(F, G_{mn\mu\nu})_{L^2(\mathbb{R}^2)}\right|^2 \\
= &\sum_{m,n,\mu,\nu} \left|\int_{-\infty}^{\infty}\int_{-\infty}^{\infty} F(x, y)\overline{g_{mn}(x)}g_{\mu\nu}(y)\, dx\, dy\right|^2 \\
= &\sum_{m,n\in\mathbb{Z}}\sum_{\mu,\nu\in\mathbb{Z}} \left|\int_{-\infty}^{\infty}\left(\int_{-\infty}^{\infty} F(x, y)\overline{g_{mn}(x)}\, dx\right) g_{\mu\nu}(y)\, dy\right|^2 \\
= &\sum_{m,n\in\mathbb{Z}} \int_{-\infty}^{\infty}\left|\int_{-\infty}^{\infty} F(x, y)\overline{g_{mn}(x)}dx\right|^2 dy \\
= &\int_{-\infty}^{\infty}\int_{-\infty}^{\infty} |F(x, y)|^2 dx\, dy.
\end{aligned}$$

□

We can now obtain matrix representations of wavelet multipliers by means of Weyl-Heisenberg frames. Let $P_{\sigma,\varphi}$ be a wavelet multiplier with symbol σ and admissible wavelet φ satisfying the conditions of Theorem 1.3. We know that $P_{\sigma,\varphi}$ is the Hilbert-Schmidt operator A_K with kernel K given by

$$(A_K f)(x) = \int_{-\infty}^{\infty} K(x, y)\, f(y)\, dy, \quad x \in \mathbb{R},$$

for all f in $L^2(\mathbb{R})$ and

$$K(x, y) = \varphi(x)k(x - y)\overline{\varphi}(y), \quad x, y \in \mathbb{R}.$$

Write

$$K = \sum_{m,n,\mu,\nu\in\mathbb{Z}} (K, G_{mn\mu\nu})_{L^2(\mathbb{R}^2)}G_{mn\mu\nu}.$$

So,

$$P_{\sigma,\varphi} = \sum_{m,n,\mu,\nu\in\mathbb{Z}} (K, G_{m,n,\mu,\nu})_{L^2(\mathbb{R}^2)}A_{G_{mn\mu\nu}}.$$

But for all f in $L^2(\mathbb{R})$,

$$(A_{G_{mn\mu\nu}}f)(x) = \int_{-\infty}^{\infty} g_{mn}(x)\overline{g_{\mu\nu}(y)}f(y)\,dy = (f, g_{\mu\nu})_{L^2(\mathbb{R})}g_{mn}(x)$$

for all x in $\mathbb{R}$. Thus,

$$P_{\sigma,\varphi}f = \sum_{m,n,\mu,\nu\in\mathbb{Z}} (K, G_{mn\mu\nu})_{L^2(\mathbb{R}^2)}(f, g_{\mu\nu})_{L^2(\mathbb{R})}g_{mn}. \tag{4.1}$$

Hence the output $P_{\sigma,\varphi}f$ is given as a frame expansion. The wavelet multiplier $P_{\sigma,\varphi}$ is then characterized by the matrix $\left[(K, G_{mn\mu\nu})_{L^2(\mathbb{R}^2)}\right]_{m,n,\mu,\nu\in\mathbb{Z}}$.

For every positive integer N, we can construct a numerical approximation $[P_{\sigma,\varphi}f]_N$ for $P_{\sigma,\varphi}f$ by truncating the expansion (4.1) at the positive integer N, *i.e.*,

$$[P_{\sigma,\varphi}f]_N = \sum_{|m|,|n|,|\mu|,|\nu|\leq N} (K, G_{mn\mu\nu})_{L^2(\mathbb{R}^2)}(f, g_{\mu\nu})_{L^2(\mathbb{R})}g_{mn}. \tag{4.2}$$

So, the truncated wavelet multiplier $[P_{\sigma,\varphi}]_N$ at the positive integer N is represented by the matrix $\left[(K, G_{mn\mu\nu})_{L^2(\mathbb{R}^2)}\right]_{|m|,|n|,|\mu|,|\nu|\leq N}$.

5. Numerical computations of wavelet multipliers

As a test case, we look at the wavelet multiplier $P_{\sigma,\varphi}$ with symbol σ and admissible wavelet $\varphi = \chi_{[-1,1]}$. For simplicity, we use $g = \chi_{[0,1]}$ as the admissible wavelet to generate a simple Weyl-Heisenberg frame. From the locations of the supports of φ and g, (4.1) becomes

$$P_{\sigma,\varphi}f = \sum_{m\in\mathbb{Z}}\left(\sum_{\mu\in\mathbb{Z}}(K, G_{m0\mu0})_{L^2(\mathbb{R}^2)}\right)(f, g_{\mu0})_{L^2(\mathbb{R})}g_{m0}. \tag{5.1}$$

For the particular choices of φ and g, we get for $|m|, |\mu| \leq N$,

$$(K, G_{m0\mu0})_{L^2(\mathbb{R}^2)} = \int_0^1\left(\int_0^1 k(x-y)e^{2\pi imx}\,dx\right)e^{-2\pi i\mu y}\,dy. \tag{5.2}$$

In the following examples, we illustrate the numerical computations of the truncated wavelet multipliers $[P_{\sigma,\varphi}]_N$ with two different symbols σ for a positive integer N. The numerical computations are implemented in MATLAB 7.0.

Example 5.1. Let $\sigma = \chi_{[-1,1]}$. Then k is given by

$$k(x) = (2\pi)^{-1/2}\check{\sigma} = \frac{\operatorname{sinc}(x)}{\pi}, \quad x \in \mathbb{R}.$$

Substituting the k into (5.2) and carrying out the integration produce the matrix $\left[(K, G_{m0\mu0})_{L^2(\mathbb{R}^2)}\right]_{|m|,|\mu|\leq N}$ for every positive integer N. Since σ is an even function, it can be shown that the imaginary part of (5.2) is zero. For $N = 3$, the matrix $\left[(K, G_{m0\mu0})_{L^2(\mathbb{R}^2)}\right]_{|m|,|\mu|\leq 3}$ is the real 7×7 matrix

$$\begin{bmatrix} 0.000 & 0.000 & 0.001 & 0.000 & -0.001 & -0.000 & -0.000 \\ 0.000 & 0.001 & 0.001 & -0.000 & -0.001 & -0.001 & -0.000 \\ 0.001 & 0.001 & 0.003 & -0.002 & -0.003 & -0.001 & -0.001 \\ 0.000 & -0.000 & -0.002 & 0.311 & -0.002 & -0.000 & 0.000 \\ -0.001 & -0.001 & -0.003 & -0.002 & 0.003 & 0.001 & 0.001 \\ -0.000 & -0.001 & -0.001 & -0.000 & 0.001 & 0.001 & 0.000 \\ -0.000 & -0.000 & -0.001 & 0.000 & 0.001 & 0.000 & 0.000 \end{bmatrix}.$$

Example 5.2. Let $\sigma(x) = e^{-x^2/2}$, $x \in \mathbb{R}$. Then k is given by

$$k(x) = (2\pi)^{-1/2}\check{\sigma} = (2\pi)^{-1/2}e^{-x^2/2}, \quad x \in \mathbb{R}.$$

Computing (5.2) numerically for $N = 3$, the matrix $\left[(K, G_{m0\mu0})_{L^2(\mathbb{R}^2)}\right]_{|m|,|\mu|\leq 3}$ is the real 7×7 matrix

$$\begin{bmatrix} 0.001 & 0.001 & 0.003 & -0.000 & -0.003 & -0.001 & -0.001 \\ 0.001 & 0.002 & 0.004 & -0.001 & -0.004 & -0.002 & -0.001 \\ 0.003 & 0.004 & 0.010 & -0.008 & -0.009 & -0.004 & -0.003 \\ -0.000 & -0.001 & -0.008 & 0.370 & -0.008 & -0.001 & -0.000 \\ -0.003 & -0.004 & -0.009 & -0.008 & 0.010 & 0.004 & 0.003 \\ -0.001 & -0.002 & -0.004 & -0.001 & 0.004 & 0.002 & 0.001 \\ -0.001 & -0.001 & -0.003 & -0.000 & 0.003 & 0.001 & 0.001 \end{bmatrix}.$$

Note that in both examples, the matrix $\left[(K, G_{m0\mu0})_{L^2(\mathbb{R}^2)}\right]_{|m|,|\mu|\leq 3}$ is symmetric with respect to the diagonals through the center, *i.e.*, the position $m, \mu = 0$, and anti-symmetric with respect to the horizontal and vertical axes through the center in the m-μ plane. More interestingly, the value of the matrix is large at the center and decays rapidly as the magnitudes of m and μ increase. This indicates that the truncated frame expansion (4.2) already provides good approximations even for very small values of N.

References

[1] O. Christensen, An Introduction to Frames and Riesz Bases, Birkhäuser, 2003.

[2] I. Daubechies, A. Grossmann and Y. Meyer, Painless nonorthogonal expansions, J. Math. Phys. 27 (1986), 1271–1283.

[3] R.J. Duffin and A.C. Schaeffer, A class of nonharmonic Fourier series, Trans. Amer. Math. Soc. 72 (1952), 341–366.

[4] Z. He and M.W. Wong, Wavelet multipliers and signals, J. Austral. Math. Soc. Ser. B 40 (1999), 437–446.

[5] H.J. Landau and H.O. Pollak, Prolate spheroidal wave functions, Fourier analysis and uncertainty II, Bell Syst. Tech. J. 40 (1961), 65–84.

[6] H.J. Landau and H.O. Pollak, Prolate spheroidal wave functions, Fourier analysis and uncertainty III, Bell Syst. Tech. J. 41 (1962), 1295–1336.

[7] D. Slepian, On bandwidth, Proc IEEE 64 (1976), 292–300.

[8] D. Slepian, Some comments on Fourier analysis, uncertainty and modelling, SIAM Rev. 25 (1983), 379–393.

[9] D. Slepian and H.O. Pollak, Prolate spheroidal wave functions, Fourier analysis and uncertainty I, Bell Syst. Tech. J. 40 (1961), 43–64.

[10] M.W. Wong, Weyl Transforms, Springer-Verlag, 1998.

[11] M.W. Wong, An Introduction to Pseudo-Differential Operators, Second Edition, 1999.

[12] M.W. Wong, Wavelet Transforms and Localization Operators, Birkhäuser, 2002.

[13] M.W. Wong and Z. Zhang, Trace class norm inequalities for wavelet multipliers, Bull. London Math. Soc. 34 (2002), 739–744.

[14] Z. Zhang, Localization Operators and Wavelet Multipliers, Ph.D. Dissertation, York University, 2003.

M.W. Wong and Hongmei Zhu
Department of Mathematics and Statistics
York University
4700 Keele Street
Toronto, Ontario M3J 1P3, Canada
e-mail: mwwong@mathstat.yorku.ca
e-mail: hmzhu@mathstat.yorku.ca

Operator Theory:
Advances and Applications, Vol. 167, 183–190

Clifford Algebra-valued Admissible Wavelets Associated to More than 2-dimensional Euclidean Group with Dilations

Jiman Zhao and Lizhong Peng

Abstract. In this paper, we consider the Clifford algebra-valued admissible wavelets, which are associated to more than 2-dimensional Euclidean group with Dilations. We give an explicit characterization of the admissibility condition in terms of the Fourier transform, study the properties of this kind of wavelet transform, also give a family of admissible wavelets.

Mathematics Subject Classification (2000). Primary: 42B10, 44A15.

Keywords. Wavelets, Clifford algebra.

1. Introduction

There are two branches in wavelet theory: admissible (or continuous) wavelet transform and discrete orthogonal wavelets generated by multiresolution analysis. In the classical case, Grossman and Morlet ([6]) introduced the wavelet transform in the one-dimensional case, where the group G is the affine group $ax+b$. After this there are many papers about admissible wavelets and decomposition of function space associated with the group $SL(2,R)$ ([7], [8], [10]). In ([9]), Murenzi generalized this to more than one dimension.

In Clifford analysis, by using Clifford algebra, L. Andersson, B. Jawerth and M. Mitrea constructed higher-dimensional discrete orthogonal wavelets, see [1]. As for Clifford-valued admissible wavelets, in their paper (see [3], [4], [5]), Brackx and Sommen studied the radial and generalized Clifford-Hermite continuous wavelet transforms in R^n. In ([13]), quaternion-valued admissible wavelets have been characterized. In this paper, we consider the Clifford algebra-valued admissible wavelets, give an explicit characterization of the admissibility condition in terms of Fourier transform, we also give a family of admissible wavelets.

Research supported by NNSF of China No.10471002 and 973 project of China G1999075105.

Now we give an overview of some basic facts which are concerned with Clifford analysis ([2]).

The real Clifford algebra R_n over R^n is defined as:

$$R_n = \{ \sum_{I \subseteq \{1,2,\ldots,n\}} x_I e_I : x_I \in R\}$$

where $\{e_1, \ldots, e_n\}$ is an orthogonal basis of R^n, and $e_I = e_{i_1} e_{i_2} \cdots e_{i_k}$ with $I = \{i_1, i_2, \ldots, i_k\}, i_1 < i_2 < \cdots < i_k$.

The non-commutative multiplication in the Clifford algebra R_n is given by:

$$e_i e_j + e_j e_i = -2\delta_{i,j}, \quad i, j = 1, 2, \ldots, n$$

An involution on R^n is defined by

$$\overline{e}_I = \overline{e}_{i_k} \cdots \overline{e}_{i_1}, \overline{e}_{i_j} = -e_{i_j}.$$

A R_n-valued function $f(x_1, \cdots, x_n)$ is called left monogenic in an open region Ω of R^n, if:

$$Df(x) = \sum_{j=1}^{n} e_j(\frac{\partial}{\partial x_j}) f(x) = 0,$$

where $D = \sum_{j=1}^{n} e_j(\frac{\partial}{\partial x_j})$ is the Dirac operator in R^n. We call $\mathcal{M}_k$ the Clifford module of the left monogenic homogeneous polynomials of degree k, its dimension is given by

$$\dim \mathcal{M}_k = \frac{(m + k - 2)!}{(m - 2)!k!}$$

Definition *Let R_n be a real Clifford algebra. A Clifford module $L^2(R^n, R_n)$ is denoted by*

$$L^2(R^n, R_n) = \left\{ f : R^n \to R_n, f = \sum_I f_I e_I \;\middle|\; \Big(\sum_I \|f_I\|^2_{L^2(R^n)}\Big)^{\frac{1}{2}} < \infty, f_I \in L^2(R^n), \forall I \right\}.$$

The inner product of $L^2(R^n, R_n)$ is denoted by

$$(f, g) = \int_{R^n} \overline{f}(x) g(x) dx.$$

2. Clifford algebra-valued admissible wavelet transform

Let $n > 2$, think about the inhomogeneous group $IG(n) = [R_+ \times SO(n)] \otimes R^n$, the group law of $IG(n)$ is given by

$$\{v, b\}\{v', b'\} = \{vv', vb' + b\},$$

where $v = ar, a > 0, r \in SO(n), b \in R^n$. The group $IG(n)$ is a non-unimodular Lie group with: left Haar measure $a^{-(n+1)} dadrdb$.

First we define a nature unitary representation of $IG(n)$ in $L^2(R^n, R_n)$. We start by considering the operators corresponding to three types of operations that constitute $IG(n)$.

Given a function $\phi \in L^2(R^n, R_n)$, we define:

$$(T^b\phi)(x) = \phi(x-b), x, b \in R^n, \quad \text{(unitary shift operator)}$$
$$D^a\phi(x) = a^{\frac{-n}{2}}\phi(a^{-1}x), \quad \text{(unitary dilation operator)},$$
$$R^r\phi(x) = \phi(r^{-1}x), x \in R^n, \quad \text{(unitary rotation operator)}.$$

Like the classical case, we have the following commutation relations:

$$T^b D^a = D^a T^{\frac{b}{a}},$$

$$R^r D^a = D^a R^r,$$

then we construct the operator

$$U_{a,r,b} = T^b D^a R^r,$$

with action

$$U_{a,r,b}\phi(x) = a^{\frac{-n}{2}}\phi(a^{-1}r^{-1}(x-b)).$$

For convenience, we denote $U_{a,r,b}\phi(x) = \phi_{a,r,b}(x)$, then it is easy to check that $U_{a,r,b}$ is a representation of $IG(n)$ in $L^2(R^n, R_n)$.

By the Fourier transform, we have

$$\phi_{a,r,b}{}^{\wedge}(\xi) = a^{\frac{n}{2}} e^{-ib\cdot\xi}\hat{\phi}(ar^{-1}\xi).$$

Like the classical case, we can prove that this representation is irreducible.

Now we define the operator W_ψ on $L^2(R^n, R_n)$ by

$$\begin{aligned} L^2(R^n, R_n) &\to L^2(IG(n), R_n, a^{-(n+1)}dadrdb) \\ f(x) &\to W_\psi f(a,r,b) \end{aligned}$$

where

$$W_\psi f(a,r,b) = (\psi_{a,r,b}, f)_{L^2(R^n,R_n)}. \tag{1}$$

By direct calculation, we have

$$W_\psi f(a,r,b) = a^{\frac{n}{2}}(\overline{\hat{\psi}(ar^{-1}(\cdot))}\hat{f}(\cdot))^{\wedge}(-b). \tag{2}$$

The inner product of $L^2(IG(n), a^{-(n+1)}dadrdb)$ is defined as follows:

$$\langle f, g\rangle_{L^2(IG(n),R_n)} = \int_{IG(n)} \overline{f(a,r,b)}g(a,r,b)a^{-(n+1)}dadrdb, \tag{3}$$

then the space $L^2(IG(n), a^{-(n+1)}dadrdb)$ becomes a right Hilbert R_n-module.

In order to get the reproducing formula like in the classical case, we calculate the $\langle W_{psi} f, W_\psi g\rangle_{L^2(IG(n),R_n)}$. Using (1), (2), (3), we have

$$\begin{aligned}
\langle W_\psi f, W_\psi g\rangle &= \int_0^{+\infty}\int_{SO(n)}\int_{R^n} \overline{W_\psi f}(a,r,b) W_\psi g(a,r,b) a^{-(n+1)} dadrdb \\
&= \int_0^{+\infty}\int_{SO(n)}\int_{R^n} \overline{(W_\psi f)}^\wedge (a,r,\xi)(W_\psi g)^\wedge (a,r,\xi) a^{-(n+1)} dadrd\xi \\
&= \int_0^{+\infty}\int_{SO(n)}\int_{R^n} a^n \overline{\hat{f}}\hat{\psi}(ar^{-1}(\xi))\overline{\hat{\psi}}(ar^{-1}(\xi))\hat{g}(\xi) a^{-(n+1)} dadrd\xi \\
&= \int_{R^n} \overline{\hat{f}}(\xi)(\int_0^{+\infty}\int_{SO(n)} \hat{\psi}(ar^{-1}(\xi))\overline{\hat{\psi}}(ar^{-1}(\xi)) a^{-1} dadr)\hat{g}(\xi)d\xi \\
&= \int_{R^n} \overline{\hat{f}}(\xi)(\int_0^{+\infty}\int_{SO(n)} \hat{\psi}(a\eta)\overline{\hat{\psi}}(a\eta) a^{-1} dad\eta)\hat{g}(\xi)d\xi \\
&= \int_{R^n} \overline{\hat{f}}(\xi)\int_{R^n} \hat{\psi}(a\eta)\overline{\hat{\psi}}(a\eta)|\eta|^{-n} d\eta\hat{g}(\xi)d\xi
\end{aligned} \tag{4}$$

On the other hand, by (1), (3), we have

$$\begin{aligned}
\langle W_\psi f, W_\psi g)\rangle &= \int_0^{+\infty}\int_{SO(n)}\int_{R^n} \overline{W_\psi f}(a,r,b) W_\psi g(a,r,b) a^{-(n+1)} dadrdb \\
&= \int_{IG(n)} \overline{W_\psi f}(a,r,b)\int_{R^n} \overline{\psi}_{a,r,b}(x) g(x) dx a^{-(n+1)} dadrdb \\
&= \int_{R^n} (\int_{IG(n)} \overline{W_\psi f}(a,r,b)\overline{\psi}_{a,r,b}(x) a^{-(n+1)} dadrdb) g(x)dx \\
&= (\int_{IG(n)} \psi_{a,r,b}(\cdot) W_\psi f(a,r,b) a^{-(n+1)} dadrdb, g).
\end{aligned} \tag{5}$$

Denote

$$C_\psi = \int_{R^n} \frac{\hat{\psi}(\xi)\overline{\hat{\psi}}(\xi)}{|\xi|^n} d\xi,$$

then by (4) and (5), we get

$$\langle W_\psi f, W_\psi g\rangle = \int_{R^n} \overline{C_\psi \hat{f}(\xi)}\hat{g}(\xi)d\xi = (C_\psi f, g),$$

thus

$$C_\psi f(x) = \int_{IG(n)} \psi_{a,r,b}(x) W_\psi f(a,r,b) a^{-(n+1)} dadrdb$$

holds in the weak sense.

So the reproducing formula holds if and only if C_ψ is the inverse Clifford number.

We call

$$C_\psi = \int_{R^n} \frac{\hat{\psi}(\xi)\overline{\hat{\psi}}(\xi)}{|\xi|^n} d\xi \text{ is the inverse Clifford number,} \tag{6}$$

the admissible condition, and call ψ the admissible wavelet if it satisfies the admissible condition (6), call the corresponding transform W_ψ admissible wavelet transform.

Denote

$$AW = \{\psi \in L^2(R^n, R_n), C_\psi = \int_{R^n} \frac{\hat{\psi}(\xi)\overline{\hat{\psi}}(\xi)}{|\xi|^n} d\xi \text{ is the inverse Clifford number,}\} \tag{7}$$

and denote an "inner product" in AW as follows:

$$\langle \psi, \varphi \rangle_{AW} = \int_{R^n} \frac{\hat{\psi}(\xi)\overline{\hat{\varphi}}(\xi)}{|\xi|^n} d\xi.$$

The AW here is different from the one in [13], there the AW is a quaternion module, while in this paper the AW is not a linear space.

The wavelet transform has the following properties: From (4) and (5), we can get the following theorem.

Theorem 1. *For $f(x), g(x) \in L^2(R^n, R_n)$, and $\psi, \varphi \in AW$,*

$$\langle W_\psi f, W_\varphi g \rangle_{L^2(IG(n), R^n, a^{-(n+1)} dadrdb)} = (\langle \psi, \varphi \rangle\rangle_{AW} f, g)_{L^2(R^n, R_n)}.$$

From Theorem 1, we know the continuous wavelet transform is an isometry, and we also get the reproducing formula.

Theorem 2. (*Reproducing formula*) *Let $f(x) \in L^2(R^n, R_n)$, $\psi \in AW$, then*

$$f(x) = C_\psi^{-1} \int_0^{+\infty} \int_{SO(n)} \int_{R^n} \psi_{a,r,b} W_\psi f(a, r, b) \psi(x) a^{-(n+1)} dadrdb.$$

Proof. By (4) and (5), we can get the conclusion in the weak sense.

Denote $A_\psi = \{W_\psi f(x) : f \in L^2(R^n, R_n), \psi \in AW\}$, then A_ψ is a Hilbert space with reproducing kernel.

Theorem 3. (*Reproducing kernel*)

Let $K(a, r, a', r')$ be the reproducing kernel of A_ψ, then

$$K(a, r, b, a', r', b) = c_\psi^{-1}(\psi_{a',r',b'}, \psi_{a,rb}).$$

Proof. From Theorem 2, we get

$$\begin{aligned} &W_\psi f(a', r', b') \\ &= c_\psi^{-1} \textstyle\int_{R^n} (\int_{IG(n)} \overline{\psi}_{a',r',b'}(x) \psi_{a,r,b}(x) W_\psi f(a, r, b) a^{-(n+1)} dadrdb) dx \\ &= c_\psi^{-1} \textstyle\int_{IG(n)} (\int_{R^n} \overline{\psi}_{a',r',b'}(x) \psi_{a,r,b}(x) dx) W_\psi f(a, r, b) a^{-(n+1)} dadrdb. \end{aligned}$$

So we have the reproducing kernel

$$K(a, r, b, a', r', b) = c_\psi^{-1}(\psi_{a',r',b'}, \psi_{a,rb}).$$

3. Examples of Clifford algebra-valued admissible wavelets

In this part we give the examples of admissible wavelets.

We know from ([11]) that the orthogonal basis of $L^2(R^n, R_n)$ is

$$\{\varphi^{(m,k,j)} = H_{m,k}(x)P_k^{(j)}(x)e^{\frac{x^2}{4}}, m, k \in N, j \leq \dim \mathcal{M}_k\},$$

where $\{P_k^{(j)}(x)\}$ is the basis of $\mathcal{M}_k$, i.e.,

$$H_{2l,k}(x) = (-2)^l l! L_l^{k+\frac{n}{2}-1}(\frac{|x|^2}{2}), \text{ when } m = 2l,$$

$$H_{2l+1,k}(x) = (-2)^l l! x L_l^{k+\frac{n}{2}}(\frac{|x|^2}{2}), \text{ when } m = 2l+1.$$

So

$$\int_{R^n} \varphi^{(m,k,j)}(x)\overline{\varphi}^{(m',k',j')}(x)dx = \delta_{m,m'}\delta_{k,k'}\delta_{j,j'}.$$

Thus

$$\int_{R^n} |\varphi^{(m,k,j)}(x)|^2 dx = \int_{R^n} |\hat{\varphi}^{(m,k,j)}(\xi)|^2 d\xi$$
$$= \int_{R^n} |\hat{\varphi}^{(m,k,j)}(\xi)|^2 |\xi|^n |\xi|^{-n} d\xi = 1 < +\infty.$$

Let $\psi^{(m,k,j)}(x)$ satisfy

$$|\hat{\psi}^{(m,k,j)}(\xi)| = |\hat{\varphi}^{(m,k,j)}(\xi)||\xi|^{\frac{n}{2}}$$

then $\psi^{(m,k,j)}(x)$ are admissible wavelets.

When n is even, let

$$\hat{\psi}^{(m,k,j)}(\xi) = \overline{\xi^{\frac{n}{2}}\hat{\varphi}^{(m,k,j)}(\xi)},$$

then

$$\psi^{(m,k,j)}(x) = \overline{(-i)^{\frac{n}{2}} D^{\frac{n}{2}} \varphi^{(m,k,j)}(x)}$$

and they satisfy the following formula:

$$\int_{R^n} \hat{\psi}^{(m,k,j)} \overline{\hat{\psi}}^{(m',k',j')}(\xi)|\xi|^{-n} d\xi$$
$$= \int_{R^n} \overline{\hat{\varphi}}^{(m,k,j)} \overline{\xi}^{\frac{n}{2}} \xi^{\frac{n}{2}} \hat{\varphi}^{(m',k',j')}(\xi)|\xi|^{-n} d\xi$$
$$\int_{R^n} \overline{\hat{\varphi}}^{(m,k,j)} \hat{\varphi}^{(m',k',j')}(\xi) d\xi$$
$$= \delta_{m,m'}\delta_{k,k'}\delta_{j,j'}.$$

When n is odd, let

$$\hat{\psi}^{(m,k,j)}(\xi) = |\xi|^{-\frac{1}{2}} \overline{\xi^{\frac{n+1}{2}} \hat{\varphi}^{(m,k,j)}(\xi)} = \overline{|\xi|^{-\frac{1}{2}} - i^{\frac{n+1}{2}} (D^{\frac{n+1}{2}} \varphi^{(m,k,j)})^{\wedge}(\xi)}$$

then

$$\psi^{(m,k,j)}(x) = C I_{\frac{1}{2}} \overline{(-i^{\frac{n+1}{2}} D^{\frac{n+1}{2}} \varphi^{(m,k,j)})(x)},$$

where C is a real number,

$$I_\alpha(f)(x) = \pi^{\frac{n}{2}} 2^\alpha \Gamma(\frac{\alpha}{2})\Gamma(\frac{n-\alpha}{2})^{-1} \int_{R^n} |x-y|^{-n+\alpha} f(y)$$

is the Riesz potential ([12]).

In this case, $\{\psi^{(m,k,j)}$ also satisfy the following formula.

$$\int_{R^n} \hat{\psi}^{(m,k,j)} \overline{\hat{\psi}}^{(m',k',j')}(\xi)|\xi|^{-n} d\xi = \delta_{m,m'}\delta_{k,k'}\delta_{j,j'}$$

Denote

$$A_{m,k,j} = \{W_{\psi^{(m,k,j)}} f(a,r,b) : f \in L^2(R^n, R_n)\}.$$

We have the following property.

Theorem 4. *For* $W_{\psi^{(n,k,i)}} f(a,r,b) \in A_{m,k,i}$ *and* $W_{\psi^{(m',k',i')}} g(a,r,b) \in A_{m',k',i'}$,

$$\langle W_{\psi^{(m,k,i)}} f, W_{\psi^{(m',k',i')}} g\rangle_{L^2(IG(n),R^n,a^{-(n+1)}dadrdb)} = \delta_{m,m'}\delta_{k,k'}\delta_{i,i'}(f,g)_{L^2(R^n,R_n)}$$

Proof. We use Theorem 1 to prove the result. Both $A_{m,k,i}$ and $A_{m',k',i'}$ are orthogonal subspaces of $L^2(IG(n), R_n, a^{-(n+1)}dadrdb)$. But we cannot get the orthogonal decomposition of $L^2(IG(n), R^n, a^{-(n+1)}dadrdb)$ (like in [14]), since the group $IG(n)$ is large in this case ($n > 2$).

Acknowledgement

The authors wish to thank Professor Heping Liu for giving useful information.

References

[1] L. Andersson, B. Jawerth and M.Mitrea, *The Cauchy singular integral operator and Clifford wavelets,* in: Wavelets: mathematics and applications, CRC Press, J.J. Benedetto and M.W. Frazier (eds.), 1994, 525–546.

[2] F. Brackx, R. Delanghe, F. Sommen, Clifford analysis, Pitman Publ., 1982.

[3] F. Brackx, F. Sommen, *Clifford-Hermite Wavelets in Euclidean Space,* J. Fourier Anal. Appl., 6(3)(2000), pp. 209–310.

[4] F. Brackx, F. Sommen, *The continuous wavelet transform in Clifford analysis,* in Clifford Analysis and Its Applications, NATO ARW Series, F.Brackx, JSR. Chisholm, V.Soucek eds., Kluwer Acad. Publ., Dordrecht, The Netherlands, 2001, pp. 9–26.

[5] F. Brackx, F. Sommen, *The generalized Clifford-Hermite continuous wavelet transform,* Advances in Applied Clifford Algebras, 11 (2001), pp. 219–231.(Proceedings of the Cetraro conference in October 1998).

[6] A. Grossman, J. Morlet, Decomposition of Hardy functions into square integrable wavelets of constant shape, *SIAM J. Math. Anal.,* **15** (1984), 723–736.

[7] Q.T. Jiang, L.Z. Peng, Wavelet transform and Toeplitz-Hankel type operators, *Math. Scand.,* **70** (1992), 247–264.

[8] Q.T. Jiang L.Z.Peng, Toeplitz and Hankel type operators on the upper half-plane, *Integral Eqns. Op. Th.,* **15** (1992), 744–767.

[9] R. Murenzi, Wavelet Transforms Associated to the n-dimensional Euclidean Group with Dilations, in *Wavelet, Time-Frequency Methods and Phase Space* (ed. Combes. J.), 1989, Boston-London: Jones and Bartlett Publishers.

[10] T. Paul. Functions analytic on the half plane as quantum mechanical states, *J. Math. Phys.*, **25** (1984), 3252–3263.

[11] F. Sommen, Special functions in Clifford analysis and Axial Symmetry, *Journal of Mathematical and Applications,* **130,** 110–133 (1988).

[12] E.M. Stein, G. Weiss, On theory of H^p spaces, Acta Math., 103, 1960.

[13] J.M. Zhao, L.Z. Peng, *Quaternion-valued admissible Wavelets Associated with the 2-dimensional Euclidean Group with Dilations,* J. Nat. Geom., 20 (2001), pp. 21–32.

[14] J.M. Zhao, L.Z. Peng, *Quaternion-valued Admissible Wavelets and Orthogonal Decomposition of* $L^2(IG(2), H)$, submitted.

Jiman Zhao
School of Mathematical Sciences
Beijing Normal University
Beijing 100875, P.R. China
e-mail: jzhao@pku.org.cn

Lizhong Peng
LMAM, School of Mathematical Sciences
Peking University
Beijing 100080, P.R.China
e-mail: lzpeng@pku.edu.cn

Operator Theory: Advances and Applications

OT 155: Ashino, R. / Boggiatto, P. / Wong, M.W. (Eds.), Advances in Pseudo-Differential Operators (2004). ISBN 3-7643-7140-4

OT 154: Janas, J. / Kurasov, P. / Naboko, S. (Eds.), Spectral Methods for Operators of Mathematical Physics (2004). ISBN 3-7643-7133-1

OT 153: Gaspar, D. / Gohberg, I. / Timotin, D. / Vasilescu, F.H. / Zsido, L. (Eds.), Recent Advances in Operator Theory, Operator Algebras, and their Applications (2004). ISBN 3-7643-7127-7

OT 152: Eidelman, S.D. / Ivasyshen, S.D. / Kochubei, A.N., Analytic Methods in the Theory of Differential and Pseudo-differential Equations of Parabolic Type (2004). ISBN 3-7643-7115-3

OT 151: Gil, J.B. / Krainer, T. / Witt, I. (Eds.), Aspects of Boundary Problems in Analysis and Geometry (2004). Subseries **A**dvances in **P**artial **D**ifferential **E**quations ISBN 3-7643-7069-6

OT 150: Rabinovich, V. / Roch, S. / Silbermann, B., Limit Operators and their Applications in Operator Theory (2004). ISBN 3-7643-7081-5

OT 149: Ball, J.A. / Helton, J.W. / Klaus, M. / Rodman, L. (Eds.), Current Trends in Operator Theory and its Applications (2004). ISBN 3-7643-7067-X

OT 148: Ashyralyev, A. / Sobolevskii, P.E., New Difference Schemes for Partial Differential Equations (2004). ISBN 3-7643-7054-8

OT 147: Gohberg, I. / dos Santos, A.F. / Speck, F.-O. / Teixeira, F.S. / Wendland, W. (Eds.), Operator Theoretical Methods and Applications to Mathematical Physics. The Erhard Meister Memorial Volume (2004). ISBN 3-7643-6634-6

OT 146: Kopachevsky, N.D. / Krein, S.G., Operator Approach to Linear Problems of Hydrodynamics. Volume 2: Nonself-adjoint Problems for Viscous Fluids (2003). ISBN 3-7643-2190-3

OT 145: Albeverio, S. / Demuth, M. / Schrohe, E. / Schulze, B.-W. (Eds.), Nonlinear Hyperbolic Equations, Spectral Theory, and Wavelet Transformations (2003). Subseries **A**dvances in **P**artial **D**ifferential **E**quations ISBN 3-7643-2168-7

OT 144: Belitskii, G. / Tkachenko, V., One-dimensional Functional Equations (2003). ISBN 3-7643-0084-1

OT 143: Alpay, D. (Ed.), Reproducing Kernel Spaces and Applications (2003). ISBN 3-7643-0068-X

OT 142: Böttcher, A. / dos Santos, A.F. / Kaashoek, M.A. / Brites Lebre, A. / Speck, F.-O. (Eds.), Singular Integral Operators, Factorization and Applications (2003). ISBN 3-7643-6947-7

OT 141: dos Santos, A.F. / Gohberg, I. / Manojlovic, N. (Eds.). Factorization and Integrable Systems. Proceedings of the Summer School, Faro, Portugal, 2000 (2003). ISBN 3-7643-6938-8

OT 140: Ellis, R. / Gohberg, I. Orthogonal Systems and Convolution Operators (2002). ISBN 3-7643-6929-9

OT 139: Müller, V. Spectral Theory of Linear Operators and Spectral Systems in Banach Algebras (2003). ISBN 3-7643-6912-4

OT 138: Albeverio, S. / Demuth, M. / Schrohe, E. / Schulze, B.-W. (Eds.). Parabolicity, Volterra Calculus, and Conical Singularities (2002). Subseries **A**dvances in **P**artial **D**ifferential **E**quations. ISBN 3-7643-6906-X

OT 137: Dybin, V. / Grudsky, S.M. Introduction to the Theory of Toeplitz Operators with Infinite Index (2002) ISBN 3-7643-6906-X

OT 136: Wong, M.W. Wavelet Transforms and Localization Operators (2002). ISBN 3-7643-6789-X

OT 135: Böttcher, A. / Gohberg, I. / Junghanns, P. (Eds.). Toeplitz Matrices, Convolution Operators, and Integral Equations. The Bernd Silbermann Anniversary Volume (2002) ISBN 3-7643-6877-2

OT 134: Alpay, D. / Gohberg, I. / Vinnikov, V. (Eds.). Interpolation Theory, Systems Theory and Related Topics. The Harry Dym Anniversary Volume (2002). ISBN 3-7643-6762-8

OT 133: Krall, A.M. Hilbert Space, Boundary Value Problems and Orthogonal Polynomials (2002). ISBN 3-7643-6701-6

OT 132: Albeverio, S. / Elander, N. / Everitt, W.N. / Kurasov, P. (Eds.). Operator Methods in Ordinary and Partial Differential Equations. S. Kovalevsky Symposium, University of Stockholm, June 2000 (2002). ISBN 3-7643-6790-3

OT 131: Böttcher, A. / Karlovich, Y.I. / Spitkovsky, I.M. Convolution Operators and Factorization of Almost Periodic Matrix Functions (2002). ISBN 3-7643-6672-9

OT 130: Gohberg, I. / Langer, H. (Eds.). Linear Operators and Matrices (2001). ISBN 3-7643-6655-9

OT 129: Borichev, A.A. / Nikolski, N.K. (Eds.). Systems, Approximation, Singular Integral Operators, and Related Topics (2001). ISBN 3-7643-6645-1

OT 128: Kopachevsky, N.D. / Krein, S.G. Operator Approach to Linear Problems of Hydrodynamics. Volume 1: Self-adjoint Problems for an Ideal Fluid (2001). ISBN 3-7643-5406-2

OT 127: Kérchy, L. / Foias, C.I. / Gohberg, I. / Langer, H. (Eds.). Recent Advances in Operator Theory and Related Topics (2001). ISBN 3-7643-6607-9

OT 126: Demuth, M. / Schulze, B.-W. (Eds.). Partial Differential Equations and Spectral Theory (2001). ISBN 3-7643-6219-7